T0187021

WJEC Vocational Award
ENGINEERING LEVEL 1/2

Matthew Wrigley

Published in 2019 by Illuminate Publishing Limited, an imprint of Hodder Education, an
Hachette UK Company, Carmelite House, 50 Victoria Embankment, London EC4Y 0DZ

Orders: please contact Hachette UK Distribution, Hely Hutchinson Centre, Milton Road, Didcot, Oxfordshire, OX11 7HH.
Telephone: +44 (0)1235 827827. Email: education@hachette.co.uk. Lines are open from 9 a.m. to 5 p.m., Monday to Friday.
You can also order through our website: www.hoddereducation.co.uk

British Library Cataloguing-in-Publication Data

A catalogue record for this book is available from the British Library

ISBN 978-1-912820-15-3

Printed in the UK

Impression 4
Year 2023

Hachette UK's policy is to use papers that are natural, renewable and recyclable products and made from
wood grown in well-managed forests and other controlled sources. The logging and manufacturing processes
are expected to conform to the environment regulations of the country of origin.

Every effort has been made to contact copyright holders of material reproduced in this
book. If notified, the publishers will be pleased to rectify any errors or omissions at the
earliest opportunity.

Editor: Dawn Booth
Text design and layout: Kamae Design
Cover design: Kamae Design
Cover photo: pkproject/shutterstock

WJEC learner assignment briefs and mark schemes are reproduced by permission from WJEC.

Contents

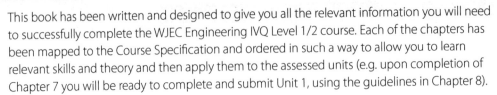

Introduction

This book has been written and designed to give you all the relevant information you will need to successfully complete the WJEC Engineering IVQ Level 1/2 course. Each of the chapters has been mapped to the Course Specification and ordered in such a way to allow you to learn relevant skills and theory and then apply them to the assessed units (e.g. upon completion of Chapter 7 you will be ready to complete and submit Unit 1, using the guidelines in Chapter 8).

This book will introduce you to many basic engineering skills and principles and will allow you to grasp a good understanding of the subject area. You will be learning how to communicate effectively as an Engineer by way of three-dimensional (3D) drawing techniques and technical drawings, as well as being able to use and identify many tools, machines and pieces of equipment that are commonplace in the engineering world.

What you will find in this book

While using this book you will come across several features that would benefit your knowledge and understanding of the subject area. These include:

Key term: a key word or phrase associated with the subject area and is a specific technical term used to demonstrate your knowledge of the vocabulary used by Engineers.

Top tip: some quick-fire advice to help you complete or understand the current task.

Task: a small task or mini-project designed to embed the knowledge you have just learnt. Also allows you to 'have a go' at applying the theory.

Chapters 11 and 12 cover a lot of the technical knowledge needed for the successful completion of Unit 2 (even though you will be tested on this information as part of Unit 3, the examination).

The route to success

To learn all the relevant skills to a standard where the top performance bands and higher grades become available to you, ideally you need access to some specialist equipment.

CAD: computer-aided design is not needed to complete the course but does produce quality outcomes. CAD is used extensively in the engineering industries and is a skill that needs to be learnt at some stage when training to be an Engineer. There are many CAD packages that, as a student, you can download free for trial periods, as well as some full CAD packages that are on offer from some companies.

Workshop: you will need access to a workshop environment to apply your learnt skills and knowledge. One of the fundamentals of engineering is the ability to manipulate materials by using processes and equipment that can only be found in a workshop, such as milling, turning and drilling. You can use hand tools or machinery to create prototypes, but understanding how to set up and use machinery safely and effectively is a fundamental skill of this course.

Course structure

The WJEC Level 1/2 Award in Engineering is a vocational course that can be started in Year 9 and run over three years or started in Year 10 and run over two years. The students completing the course will be assessed with three units and graded from Distinction Star to Level 1 Pass. The grades awarded for each unit will be assessed from the Specification Mark Scheme, where you will find the Assessment Criteria and relevant Performance Bands. You can access this information by looking at the WJEC website and selecting the Engineering Qualification: **https://www.wjec.co.uk**.

Below you will find examples on how the course will be graded and assessed.

Grading

The grading structure for the course is as follows:

Grade	Written as	GCSE equivalent
Distinction Star (Level 2)	D*	A*
Distinction (Level 2)	D	A
Merit (Level 2)	M	B
Pass (Level 2)	L2P	C
Pass (Level 1)	L1P	D

Units

The following units will be assessed:

Unit title (number)	Assessment	Content
1 Engineering Design (9791)	Internal (school/college)	6–7 page portfolio task
2 Producing Engineering Products (9792)	Internal (school/college)	Workshop making task
3 Solving Engineering Problems (9793)	External (WJEC moderators/examiners)	Exam

How will you be assessed?

You will be assessed with three units:

Unit 1: Internally assessed (guided learning hours 30)

As a guideline, Unit 1 is a 'design'-based task that requires you to submit a portfolio of approximately six–seven pages of A3. However, if you want to submit on A4 or even digitally then that is also acceptable, as it depends on your preference or access to equipment. This will be assessed/marked by your tutor.

This unit is a Controlled Assessment Task that should have a 'medium' control level, with your tutor supervising the work.

You will have seven hours to produce the assessable work.

Unit 1 could include:

- product analysis (ACCESSFM/reverse engineering, above and below analysis)
- generation of ideas (with annotation)
- development of idea/s (with links to current/existing products)
- orthographic projection (3rd angle)
- bespoke brief
- design specification
- final design
- CAD
- isometric view.

All candidates' submissions would be guided by the centre (school/college).

Unit 2: Internally assessed (guided learning hours 60)

Unit 2 expects you to demonstrate your learnt workshop skills and technical drawing skills to translate a 3rd angle orthographic projection (engineering drawing) and produce a prototype in the centre's workshop environment. You will also have to produce supporting documentation showing your understanding of the task and evidence of your working practice. This will be assessed/marked by your tutor.

This unit is a Controlled Assessment Task that should have a 'medium' control level, with your tutor supervising the work.

You will have 12 hours to produce the assessable work.

Unit 2 could include:
- manufactured prototype created in a workshop environment
- evidence of eight or nine 'skills' (e.g. milling, scribing, drilling, etc.)
- risk assessments
- production plan
- sequencing of processes
- job sheets/parts lists
- learner observation record
- evaluation.

All candidates' submissions would be guided by the centre (school/college).

Unit 3: Externally assessed (guided learning hours 30)

Unit 3 is an externally assessed examination that you will sit in a 'controlled' environment such as a school or college, supervised by invigilators. The examination will last 90 minutes and cover all aspects of the engineering specification, including knowledge and theory.

Unit 3 Problem Solving Examination could include testing of:
- theory
- knowledge
- mechanical problem solving
- electronic problem solving
- structural problem solving
- drawing skills
- knowledge of materials.

When should you sit the units?

Submitting work for all three units should be dictated by your school or college, and your tutor would give you notice of when work should begin and when the deadlines are imminent.

The chapters in this book cover the following:
- Chapters 1 to 7 cover enough of the specification for you to successfully complete and submit Unit 1, using the guidance in Chapter 8, with enough knowledge to access the top performance bands.
- Chapters 9 to 13 cover the remainder of the needed 'theory-based' knowledge in readiness for Unit 2 and Unit 3.

To successfully complete Unit 2 you will need workshop practice. You will need to complete several workshop-based projects that will take you through procedures on how to use tools and equipment safely and accurately. You will then be able to apply this knowledge to producing your prototype in Unit 2.

Unit 3 can be sat early if your tutor thinks you are ready. However, you can have the option of a RE-SIT the following year. The re-sit options would need to be discussed with your centre.

Grading the course

How do I work out my final grade?

When you complete each unit you will be awarded points. Your final grade will be dependent on how many points you scored for each unit. Awarded points are then added together to give you a final score. The following tables show you how many points each grade is worth for each unit and how many TOTAL points are needed for each grade. The following table shows you how many points you can earn for each unit:

Unit	Points per unit			
	Level 1 Pass	Level 2 Pass	Level 2 Merit	Level 2 Distinction
Unit 1 (9791)	1	2	3	4
Unit 2 (9792)	2	4	6	8
Unit 3 (9793)	1	2	3	4

> **Important note**: to attain a grade (e.g. Merit) for a unit then must attain a minimum of that grade for each/all Assessment Criteria. For example, if you attain 7 Merits and 1 Level 2 Pass in Unit 1, then your overall grade for Unit 1 will be a Level 2 Pass (2 points).

The following table shows you the overall award/qualification you can receive when you total your earned points:

Qualification received	Overall grading points	
WJEC Level 1 Vocational Award in Engineering 9790	Pass	4–6
WJEC Level 2 Vocational Award in Engineering 9790	Pass	7–10
	Merit	11–13
	Distinction	14–15
	Distinction *	16

Below is a handy checklist for you to go through every time you complete a chapter. Copy the checklist into your notebook and tick the boxes when they are covered and then put a tick in either the 'Poor', 'OK' or 'Good' box to check your understanding of the chapter. When looking back over this checklist while revising, you will quickly be able to see what areas you are strong in and what areas you will need to revise further.

Chapter title	Tick if covered	Understanding		
		Poor	OK	Good
1 Engineering Drawings				
2 Communicating Design Ideas				
3 Materials and Properties				
4 Identifying Features of Working Products				
5 Analysing and Designing Products Against a Brief				
6 Specifications				
7 Evaluating Design Ideas				
8 Unit 1 Submission				
9 Managing and Evaluating Production				
10 Health and Safety in the Workshop				
11 Engineering Tools and Equipment				
12 Engineering Processes				
13 Unit 2 Submission				
14 Effects of Engineering Achievements				
15 Engineering and the Environment				
16 Engineering Mathematical Techniques				

1 Engineering Drawings

In this chapter you are going to:
→ Learn about standardisation
→ Discover how to use standards when creating technical drawings
→ Learn how to create:
  - isometric drawings
  - cutaway drawings
  - exploded drawings
  - 3rd angle orthographic projections and sectional drawings.

This chapter will cover the following areas of the WJEC specification:

Unit 1 LO2 Be able to communicate design solutions	
AC2.1 Draw engineering design solutions	Draw (using British Standards): 3rd angle orthographic projection; isometric; dimensions and associated symbols – diameter, circumference, radius, height, depth, width; conventions – title block, dimension lines, extension lines, centre lines, metric units of measurement; hidden detail; scale

Unit 2 LO1 Be able to interpret engineering information	
AC1.1 Interpret engineering drawings	Interpret: symbols; conventions; information; calculations Sources: sketches; drawings; design specifications
AC1.2 Interpret engineering information	Engineering information: data charts; data sheets; job sheets; specifications; tolerances

Unit 3 LO4 Be able to solve engineering problems	
AC4.2 Convert between isometric sketches and 3rd angle orthographic projections	Convert: section views; construction lines; centre lines; hidden detail; standard conventions

Introduction

Engineers constantly create and use drawings as part of their day-to-day jobs. Drawings allow Engineers to physically 'see' that shape of a product, look at how something could be assembled, recognise the different views of a product, as well as identify important points such as dimensions, materials and hidden details.

The drawings produced by Engineers are used to manufacture products from PlayStations to aircraft to skyscrapers, all of which need to be highly accurate. Any mistake in the drawings would be transferred to the actual product during the manufacturing stage and huge amounts of time and money could be lost. This is why all modern technical drawings need to be STANDARDISED.

Drawing standards

There are several organisation across the globe that standardise the many products and services that humanity uses, such as plug sockets, paper and technical specifications for industrial processes. The two organisations' standards that UK-based Engineers have to conform to when drawing are the:

International Standardisation Organisation (ISO)

and

British Standards Institution (BSI) bsi.

Both ISO and BSI have developed a recognised format to **standardise** technical drawings. If an Engineer in the UK produces a set of technical drawings for a product ready to be manufactured in a factory in, say, China, then a set of standardised drawings would be needed. Using standardised drawing would enable the Chinese manufacturers to understand the technical information clearly and accurately. Drawings created with ISO and BSI standards are recognised throughout the world (ISO and BSI work together to create the exact same standards).

The standardisation numbers for technical drawing are:
* **ISO 128**
* **BSI 8888:2017**.

↑ A *kitemark*.

Key term

Kitemark: awarded by BSI when a product meets its standards.

Organisation and equipment

To successfully learn the skills necessary for engineering drawing techniques, it would be helpful to have access to some equipment. Below are some suggestions:

Drawing
* 1 × Set of graphics pencils (3H to 3B) for drawing and sketching
* 1 or 2 fine line black pens for picking out lines in a drawing or sketch
* 1 × Steel rule (or a standard ruler)
* 1 × 180° protractor
* 1 × 45° set square
* 1 × 30° set square
* 1 × Compass
* 1 × Eraser
* 1 × A3 sketchbook (lower quality is OK).

Organisation
* 1 × A4 ring binder (with or without poly-pockets)
* 1 × A3 plastic portfolio.

The basics

When drawing, in engineering there are basic rules you need to follow as well as understanding how your equipment works. Here are a few tips to get you started.

Pencils and lines

——————————— 3H pencil	
——————————— 2H pencil	**Faint thin lines** → Construction lines
——————————— H pencil	
——————————— HB pencil	
——————————— B pencil	**Dark thin lines** → Weighted lines
——————————— 2B pencil	
——————————— 3B pencil	
——————————— Fine-liner (black ink)	→ Weighted lines

Key term

Construction lines: faint, thin lines that are easy to rub out.

Top tip

When constructing a drawing remember the phrase: 'light-is-right'.

When creating a drawing you will first need to construct the overall shape with construction lines. These are faint, thin lines that are easy to rub out.

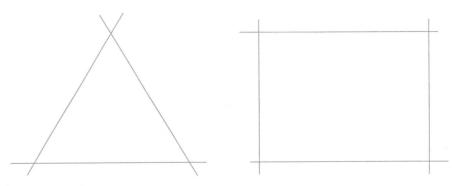

↑ A triangle (left) and rectangle (right) constructed using construction lines.

Key term

Weighted lines: define the object you are drawing making it easier to see which lines to keep and which to erase.

Task 1.1

On an A3 sheet, label the lines your pencils make. Then construct some simple shapes and pick them out using weighted lines.

When 'picking-out' the shape of your drawing you will use weighted lines. The weighted lines define the object you are drawing and make it easier to see what lines to keep and what lines to erase.

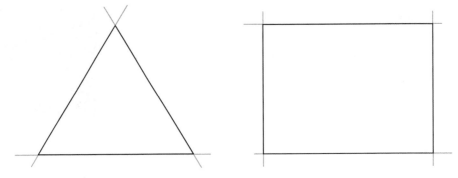

↑ A triangle (left) and rectangle (right) picked out using weighted lines.

Isometric drawing

Isometric drawing is a standardised (ISO, BSI) way of presenting designs and drawings in 3D. Isometric drawing is also a 'formal' way of presenting images in 3D and is used in many different ways to communicate information such as technical details and assembly instructions. Isometric is also the view used for many 'top-down' video games and also CAD, as the 3D view is easy to understand and navigate around.

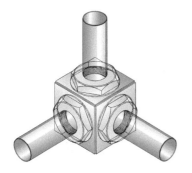

↑ *Isometric image from 'Age of Empires' video game.*

The following example has been drawn with a 30° set square; drawings are always drawn at 30° (from horizontal) in isometric projection. In isometric projection all vertical lines on an object remain vertical while all other lines are drawn at 30° to the horizontal. Isometric drawings are usually produced with drawing equipment or using CAD to ensure accuracy. When starting, you can use isometric grid paper to help.

↓ *Isometric grid that can be used to trace with.*

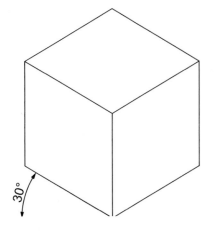

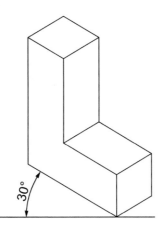

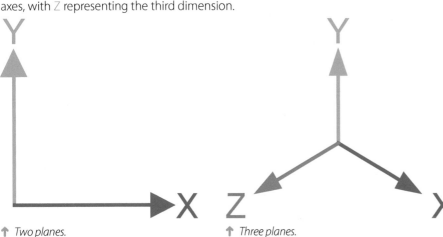

Isometric drawing is constructed using three planes. In two-dimensional (2D) drawing only two planes are used: X and Y axes. However, in 3D drawing three planes are used: X, Y and Z axes, with Z representing the third dimension.

↑ *Two planes.*

↑ *Three planes.*

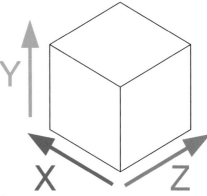

↑ *Three planes.*

Constructing isometric drawings

Now we will construct our first isometric cuboid. You will need: a pencil for CONSTRUCTION LINES, a pencil for WEIGHTED LINES, an A3 piece of paper (landscape) and a 30° set square.

Task 1.2

On an A3 sheet of paper, follow the below guide to create your first isometric crate.
1. Using your H pencil, draw a horizontal line to form your baseline.
2. Using your 30° set square, draw a vertical line from the centre of your baseline.
3. Draw a 30° line as shown.

Key term

Baseline: the horizontal line you use to 'level' your set square.

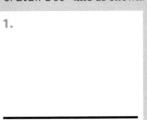

4. Reverse your set square and draw another 30° line intersecting with the first 30° line.
5. Raise your set square and draw another 30° line.
6. Reverse your set square and draw another 30° line intersecting with the previous 30° line.

Top tip

All lines on the same plane should be parallel.

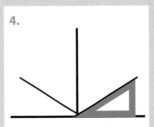

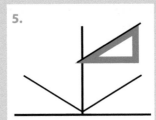

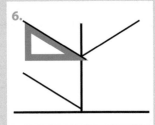

7. Rotate your set square and draw another vertical line.
8. Move your set square and draw another vertical line.
9. Rotate your set square and draw another 30° line intersecting with the vertical line.

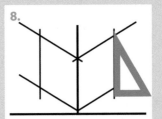

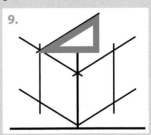

10. Reverse your set square and draw another 30° line intersecting with the previous vertical line.
11. Now you have a completed isometric cuboid in construction lines,
12. Pick-out your 3D shape with weighted lines.

Top tip

When your drawing is complete, **DO NOT** rub out the construction lines, they show the 'workings-out' ... just like mathematics.

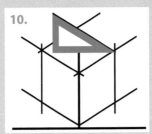

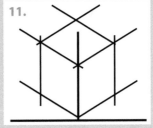

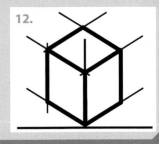

Constructing shapes in isometric

Now you know how to construct an isometric cuboid you can use that space to create other shapes. Constructing a cuboid is known as constructing a crate. Within the space of the crate you can now produce other shapes.

When artists create images they mostly build layer upon layer and understand how one affects the other to produce realistic images (e.g. understanding an animal's physiology to build the skeleton, muscles, skin and hair/fur). Engineers are more like sculptors, starting with a block of material and removing all the unwanted material to leave the shape needed (imagine the crate as a block of ice or marble). The following are some of the ways of removing material from a crate to produce a desired image.

Removing material

1. Construct a crate in isometric.

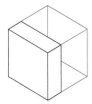

2. Trim your crate to the desired size using vertical and 30° lines.

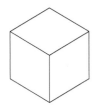

3. Identify the shape you need and remove the excess material.

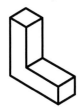

4. Pick out your shape with weighted lines and remove the construction lines.

5. Finish with shading or rendering if needed.

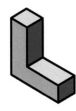

↑ *Imagine an ice sculptor carving a sculpture with a chainsaw.*

Top tip

While constructing your isometric drawing ... if it looks wrong, then it is wrong (so fix it). Trust your vision.

Extruding shapes

You can also use isometric cuboids to 'extrude' profiles through one plane to create 3D objects. Extrusions are profiles that have been extended or stretched, much like prisms.

1. Choose a face on your isometric crate and draw the desired shape. Using 30° lines 'extrude' your shape onto the back/rear face of your isometric cube.

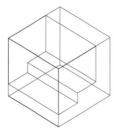

2. Join the two profiles with 30° lines to the detail points (corners, etc.).

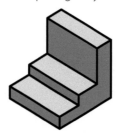

3. After picking out your shape in weighted lines, remove construction lines and finish as needed.

Key terms

Planes: the X, Y and Z axes (direction) you create in.
Extrusions: profiles that have been extended or stretched.

Task 1.3

In your sketchbooks, draw isometric drawings of the 3D shapes below.
Make sure you start this exercise by constructing isometric crates and working within the crate 'space'. Your drawings DO NOT have to be dimensionally accurate.

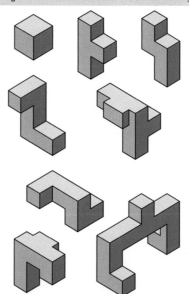

Angles and curves in isometric

Angles

Not all products we see and use are constructed from squares, flat edges and cuboids. The vast majority of items we see and use every day are made-up from angles, curves and circles, as well as squares and cuboids. Engineers need to be able to communicate ALL shapes clearly and effectively, including angles and curves.

In this section we are going to look at ANGLES and how to construct them in isometric.

Angles in one plane:

1. Construct a crate in isometric.

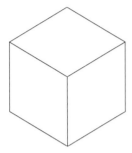

2. Choose a face, draw an angled line from one corner to the opposite corner and then extrude the angled line to the opposite face of the crate.

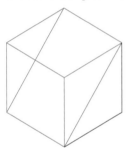

3. Pick out your angled shape using weighted lines and rub out the construction lines.

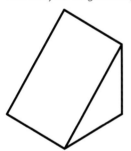

4. Finish your new angled shape with shading if needed.

Angles in two planes:

1. Construct a crate in isometric, choose a face and extrude an angled line through one plane.

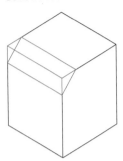

2. Repeat Step 1 but extruding your angle through a second plane.

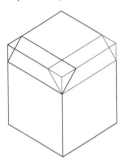

3. Repeat Steps 1 and 2 but on the base of your isometric crate.

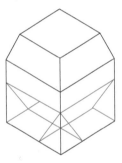

4. After picking out your shape in weighted lines, rub out the construction lines and finish/shade if needed.

Task 1.4

On an A3 sheet, construct three crates and draw shapes that include angles. Try to use all three planes for your final drawing.

Circles and curves in isometric

When drawing circles and curves in 3D you have to take into account the perspective of an object. Isometric uses something called an axonometric perspective. An axonometric perspective (isometric view) is a 'pictoral' representation of 3D objects and not a true view of how we view the world. True perspective shows lines disappearing into the distance and converging at one or more points (vanishing points) where isometric drawing shows lines running parallel (30° lines). Axonometric projections have the advantages of being clear to understand as they are used in recognised drawing formats (ISO, BSI) such as isometric, are accurate as well as allowing dimensions to be added.

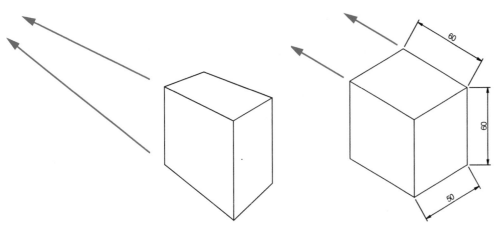

↑ *Examples of perspective. The image on the left is drawn in true perspective, with parallel lines converging to a vanishing point; the right-hand image is an example of an axonometric drawing. The lines are parallel so you can add dimensions and give the viewer more technical information. Lines converge in perspective.*

Circles in isometric drawing are known as ellipses. When looking at a cylinder (for example, a tin of beans or cola can) you know the top and bottom of that cylinder are circles. However, what you see is in fact an ellipse.

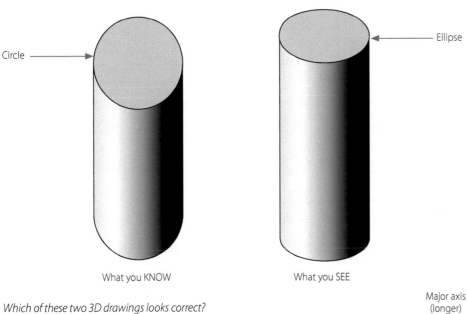

What you KNOW   What you SEE

↑ *Which of these two 3D drawings looks correct?*

When discussing ellipses we often talk about the MAJOR AXIS and the MINOR AXIS. On the right is a diagram showing what each of these is.

> **Key terms**
>
> Perspective: your point-of-view when looking at an object.
> Axonometric perspective: a pictoral representation of a 3D object that is not a true view of how you would view it.
> Vanishing points: lines that disappear into the distance.

> **Key term**
>
> Ellipse: a circle viewed in axonometric projection.

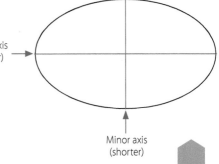

Major axis (longer)

Minor axis (shorter)

Ellipses in isometric

There are several methods of drawing ellipses accurately, using basic drawing equipment such as rulers, set squares, compasses and pencils. These methods include:

The freehand sketch method	Used when sketching-out ideas or practising your isometric drawing. Only a pencil is needed for this method.
The concentric circle method	Two concentric circles are used with the major and minor axes of the isometric 'diamond'. A compass, ruler and pencil are needed for this method.
The trammel method	Uses the major and minor axes of an isometric 'diamond' as well as a strip of paper used as a trammel. A ruler, pencil and trammel are needed for this method.
The four centre method	Uses the original circle with a series of centre points and drawn arcs to create an isometric ellipse. A ruler, compass and pencil are needed for this method.

Key term

Trammel: (or the trammel method) uses a trammel of Archimedes (also known as an ellipsograph) to draw ellipses. This method can also be replicated by using a piece of paper and a major and minor axis of the ellipse to be drawn.

Task 1.5

Using the below four centre method guide, construct an ellipse on an A3 sheet of paper.

Four centre method:
1. Draw a circle with the required diameter (for the major axis of the ellipse) within a square.
2. Quarter the circle/square.
3. Draw a line from point A to point B and from point C to point D.
4. Repeat the process but from the opposite corners.

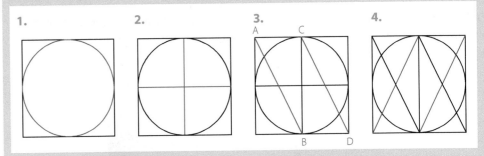

5. Using a compass, set the metal centre to point E, the pencil to point F and draw an arc between points F and G. Repeat on the opposite side.
6. Using a compass, set the metal centre to point H and draw an arc between points I and J. Repeat on the opposite side.
7. Pick out your ellipse with weighted lines. Finish.

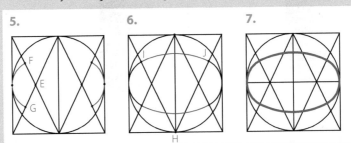

When drawing ellipses in isometric you can draw in one of the three different planes (X, Y and Z).

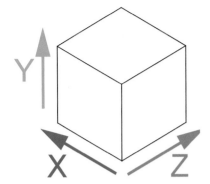

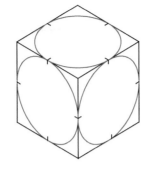

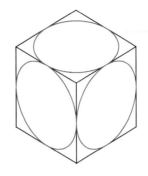

↑ *Ellipses drawn in three planes.*

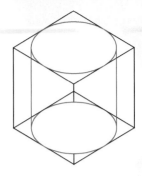

↑ *A cylinder created by drawing two ellipses in one plane.*

When sketching freehand in isometric to generate quick ideas or even discussing ideas with clients and customers, you can follow these simple rules:

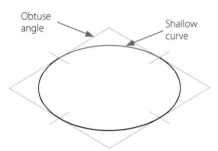

Obtuse angle

Shallow curve

↑ *Draw a SHALLOW CURVE when you have an OBTUSE ANGLE.*

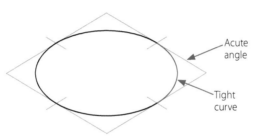

Acute angle

Tight curve

↑ *Draw a TIGHT CURVE when you have an ACUTE ANGLE.*

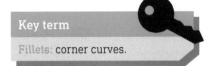

Key term

Fillets: corner curves.

Task 1.6

Drawing a trendy coffee table

Copy the drawings below into your notebook, then, using the correct equipment (set square, ruler), complete the drawing exercise, demonstrating the skills you have learnt so far. You can draw the fillets freehand.

1.

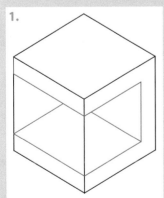

Draw an isometric crate and remove most of the middle section.

2.

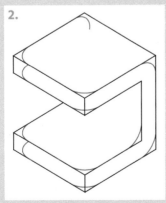

In freehand, draw sections of ellipses on the two different planes until you have created fillets on the corners and edges.

3.

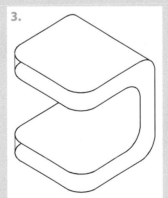

Connect the corner fillets and rub out the corner lines that have fillets.

4.

Pick-out the coffee table shape with weighted lines and shade.

Cutaway drawings

Cutaway drawings are designed to show the viewer important parts of the interior of an opaque object or product (an object that you cannot see into because it has a solid exterior/case). This is achieved by 'cutting away' parts of the exterior and leaving other parts of the exterior intact. By doing this you can show many important features of products such as the internal layout of the seating on an aeroplane, the pistons moving in an engine or how the internal components of a drill fit into the casing. Cutaway drawings also show the *different parts* of a product and how one part can interact with another.

Constructing a cutaway drawing

When constructing cutaway drawings you are trying to communicate information to the viewer that can be sometimes very complicated. By following a series of tried-and-tested guidelines, the finished drawing will not only be easier to construct but also easier to understand.

GUIDELINES
- Construct your cutaway drawing in isometric.
- When drawing your product, think of it as different parts, not a whole product.
- Only part of the exterior will be 'cut away'.
- All parts that have been 'cut' will be hatched.
- When two different parts meet, try to draw hatching in the opposite direction.

Key term

Hatching: a series of 45° parallel lines that are separated by an appropriate distance (e.g. 4mm) to show where a solid object has been cut.

Task 1.7

Find a simple object from your house (jewellery box, pen, drinking bottle, etc.) and produce an isometric cutaway drawing.

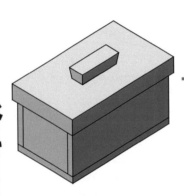

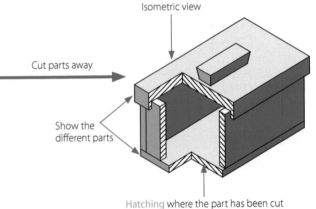

Isometric view

Cut parts away

Show the different parts

Hatching where the part has been cut

Exploded drawings

Exploded drawings are created to show all the different parts of a product and how they are assembled. You have probably used exploded drawing when using instruction manuals when assembling furniture or some toys. They are a great pictoral aid for showing how all component parts of a product interrelate.

At first glance, exploded drawings look quite complicated. In fact, they are simple isometric drawings that use only the simple drawing skills demonstrated so far in this chapter.

Key term

Assembled: put together.

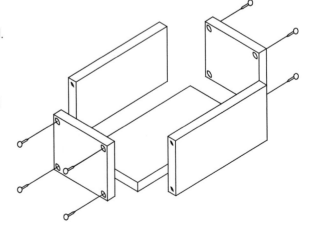

Constructing an exploded drawing

Imagine a product is exploding … then pause it in time. The result should show all the parts slightly separated. If you then slowly reverse the explosion you can see how each of those parts go together. Below you will find some useful tips on how to create an exploded drawing:

- Always draw your exploded drawing in isometric.
- Think of it as different parts, not the whole product.
- Use the same projection lines for parts that are opposite each other.

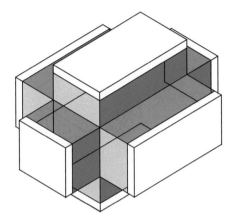

↑ *This exploded drawing has been constructed using three isometric crates. The three crates all 'intersect' in the middle. All three axis are used when projecting or extruding the crates.*

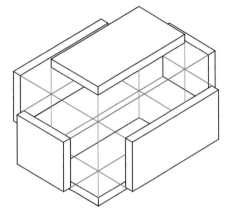

↑ *This exploded drawing has been constructed using projection lines. See how one part has been drawn and then extruded using projection lines. The parts have been projected along the X, Y and Z planes.*

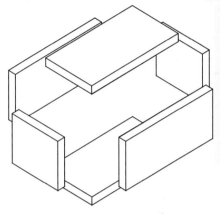

↑ *When you erase the construction lines and use weighted lines to pick out your drawing, you will end up with a completed, easy to understand exploded drawing.*

Orthographic projections

Orthographic projections are standardised drawings (ISO, BSI) that contain all the relevant technical information needed for the part/product to be made by a third party. Engineers regularly design parts and products that would be manufactured elsewhere, often in other countries such as China. The drawings therefore need to be extremely accurate with all the vital information communicated clearly and efficiently. By having standardised drawings, anyone reading the drawing would be able to understand it, as it would conform to the relevant standards from ISO and BSI. Orthographic projections are often known as technical drawings, **working drawings** or **engineering drawings** and can contain lots of relevant conventions such as:

- different views
- dimensions
- scale
- materials
- hidden detail
- centre lines
- finishes
- section views
- date the drawing was produced
- Engineer's/Designer's name
- 'angle' symbol
- title
- parts list
- manufacturing processes.

> **Key terms**
>
> Orthographic projection: (in engineering) a means of representing different views of an object by projecting it onto a plane or surface.
>
> Technical drawings: the common term used for 3rd angle orthographic projections.
>
> Conventions: technical terms.

Orthographic projections are constructed using different views of the part/product. This enables the viewer to see details that might sometimes be hidden. The views that are usually shown are:

- **front view**
- **side view**
- plan view
- sometimes an isometric or section view.

Key term

Plan view: a view from above; also known as a bird's-eye view.

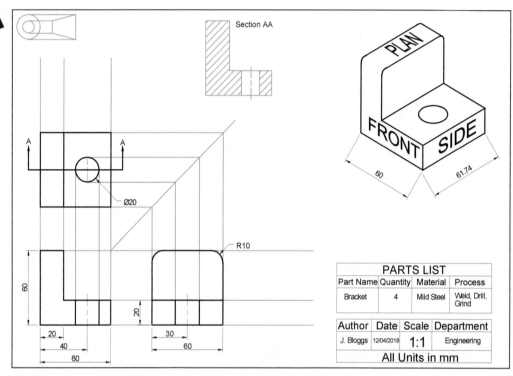

↑ *Example of an orthographic projection.*

There are two different methods of drawing in orthographic projection. These methods differ in what views you choose to show.

Below are examples of a 1st and 3rd angle orthographic projection that show how the different views of a 3D part or product are laid out.

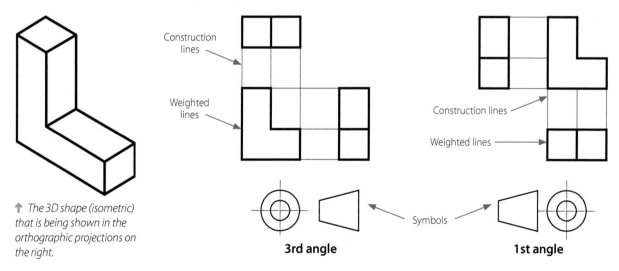

↑ *The 3D shape (isometric) that is being shown in the orthographic projections on the right.*

As well as representations of the 3D shape you can also see the 'symbols' used to show what type of drawing is being displayed (1st or 3rd angle). The symbol used is based on a 3D shape that looks like a solid lampshade or cone with the top cut off.

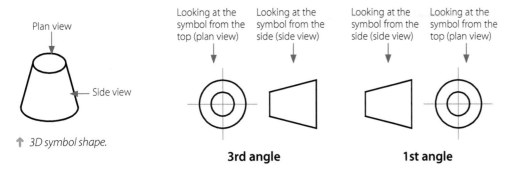

Plan view

Side view

↑ *3D symbol shape.*

Looking at the symbol from the top (plan view)

Looking at the symbol from the side (side view)

Looking at the symbol from the side (side view)

Looking at the symbol from the top (plan view)

3rd angle

1st angle

Key term

Representations: views.

Third angle orthographic projections

For this section you will be learning how to draw in 3rd angle orthographic projection.

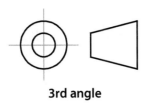

3rd angle

The way you 'rotate' the 3D object to create the views in 3rd angle orthographic projection is very specific. You must turn the object 90° in the correct direction.

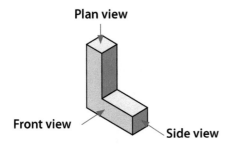

Plan view

Front view

Side view

The correct position of each view is shown below. The end result should be very accurate and be completed using the correct drawing equipment or computer aided design (CAD).

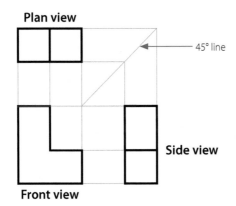

Plan view

45° line

Side view

Front view

Below is a diagram showing how the 3D object is rotated to display each of the views for 3rd angle orthographic projection:

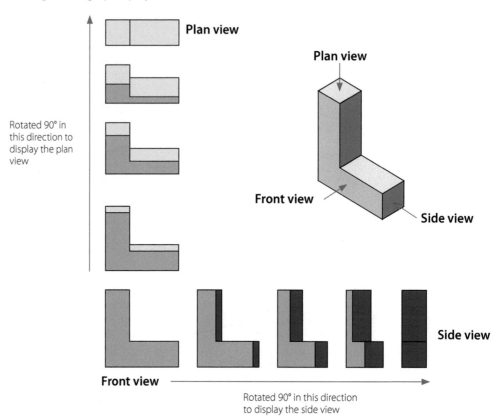

Rotated 90° in this direction to display the plan view

Plan view

Plan view

Front view

Side view

Side view

Front view

Rotated 90° in this direction to display the side view

⬆ *The 3D object showing all the views that will be displayed.*

Below is a diagram showing how the 3D object is projected onto a 'wall' to display each view for 3rd angle orthographic projection:

Task 1.8

Draw or find a few simple 3D objects and construct 3rd angle orthographic projections of them. The drawings do not have to be dimensionally accurate. Try to get the views in the correct positions (front, plan, side). Do this on an A3 sheet of paper.

Top tip

Spend some time getting used to HOW the object moves from the original (front) view, as seen in the drawings on the right. Practise drawing the different views of simple shapes.

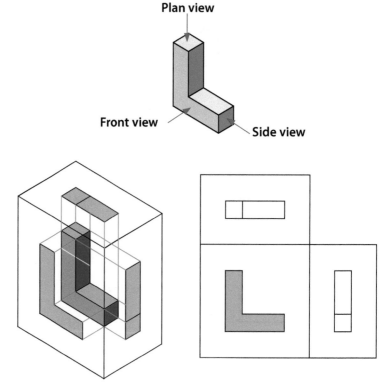

Plan view

Front view

Side view

⬆ *The 3D object showing all the views that will be displayed.*

Dimensions

Dimensioning your orthographic drawings is very important and must be completed very accurately. The size and shape of the product when it is made is dependent on the dimensions you use on your drawing. To minimise any confusion when reading an orthographic drawing, you must use a standardised way of dimensioning (BSI 8888:2017). Here are a few simple rules:

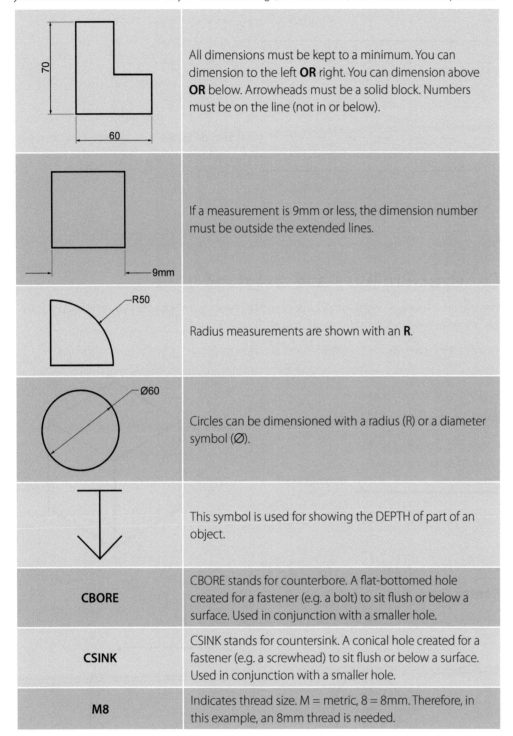

	All dimensions must be kept to a minimum. You can dimension to the left **OR** right. You can dimension above **OR** below. Arrowheads must be a solid block. Numbers must be on the line (not in or below).
	If a measurement is 9mm or less, the dimension number must be outside the extended lines.
	Radius measurements are shown with an **R**.
	Circles can be dimensioned with a radius (R) or a diameter symbol (Ø).
	This symbol is used for showing the DEPTH of part of an object.
CBORE	CBORE stands for counterbore. A flat-bottomed hole created for a fastener (e.g. a bolt) to sit flush or below a surface. Used in conjunction with a smaller hole.
CSINK	CSINK stands for countersink. A conical hole created for a fastener (e.g. a screwhead) to sit flush or below a surface. Used in conjunction with a smaller hole.
M8	Indicates thread size. M = metric, 8 = 8mm. Therefore, in this example, an 8mm thread is needed.

⬆ *Screw thread should be shown on an orthographic drawing if secrews are required.*

Lines

There are many different types of line that are used when constructing an engineering drawing. Due to the sheer amount and variety of lines used, specific lines have been created to show specific things or that have a specific job. Below are some common examples that conform to BSI 8888:2017 and what they are used for.

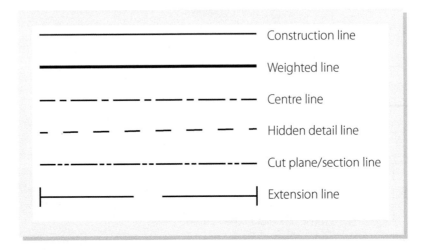

Construction lines and weighted lines

Construction lines are very thin/faint lines used to construct the shapes you are drawing. They are meant to tell you where the position of each object is.

Weighted lines are thicker/darker lines and are used to define/pick out the actual object.

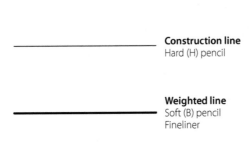

Construction line
Hard (H) pencil

Weighted line
Soft (B) pencil
Fineliner

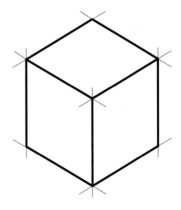

Centre lines

Centre lines are used to show the centre point of a round object.

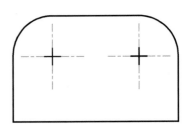

Hidden details lines

Quite often, on an orthographic drawing you will find objects that have details that will be hidden when showing a certain view (e.g. plan view, front view, side view, etc.). This hidden detail must be shown with dashed lines. Look at the drawing below to see how the hole in the object has been shown on the front and plan views with dashed lines.

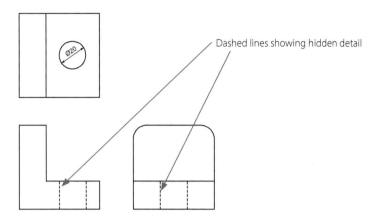

Dashed lines showing hidden detail

Section lines/cut planes

On some views (plan, front, side) you may find a section line or cut plane line. This will coincide with a 'sectional drawing' on the same page.

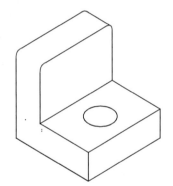

⬆ *Isometric view of object.*

Plan view

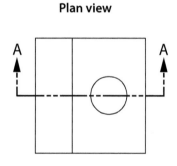

⬆ *Section line; the direction of the arrows shows what part of the object is being shown in the section drawing.*

Section view

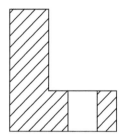

⬆ *Section drawing showing the part that is displayed. The hatching shows where the part has been cut.*

Extension lines

The extension lines are those used in dimensioning and they define the area that is being dimensioned. There should be a small, consistent gap between the object you are dimensioning and the extension lines.

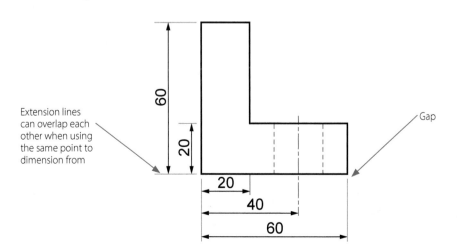

Extension lines can overlap each other when using the same point to dimension from

Gap

Section drawings

Section drawings show a product as if it had been sliced or sectioned, so you can view the interior (sometimes they are called **cross-sections**). The position of the imaginary cut is called a **section plane** or **cut plane** and is drawn with long and short dashes.

The parts of the product that have been *sectioned* will show where they have been cut by the use of **hatching**.

Hatching rules: when two different parts of the product meet in a section view the hatching must (try) to go in opposite directions. All hatching must be evenly spaced (approximately 5mm) and be at 45°.

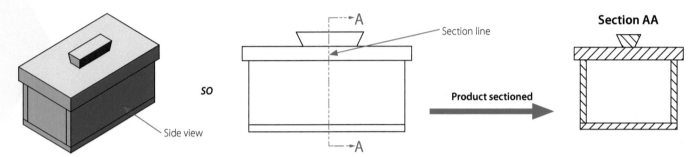

↑ *Side view of an orthographic projection.*

Task 1.9

Draw a simple 3rd angle orthographic projection. Choose a 'view' and section it using section lines. Then draw the **section view**.

Border, parts list and title block

Usually, a border and title block are drawn before the orthographic drawing is even started. This helps to keep the work neat and tidy right from the beginning. It also provides a space for important information such as your name, the title and the date. The parts list allows you to number each part of the drawing and list all the important pieces of information such as materials, amounts, sizes and finishes.

PART NUMBER	PART QUANTITY	MATERIAL	FINISH	PROCESS

TITLE	SCALE	NAME	ALL MEASUREMENTS IN mm

Projection lines and scale

Every orthographic drawing has projection lines. These projection lines are where you **project** the original view. They are very faint lines that help you to construct the drawing.

Each drawing also shows what scale it is drawn in. It is impractical to draw everything the size it actually is, so often you have to scale down or up a drawing (the drawing below is half the actual size 1:2). For instance, if your drawing is exactly the same size as the original item, this is written as 1:1. However, if your drawing is twice the size of the original, this is written as 2:1. If your drawing is halving all the measurements, this is written as 1:2.

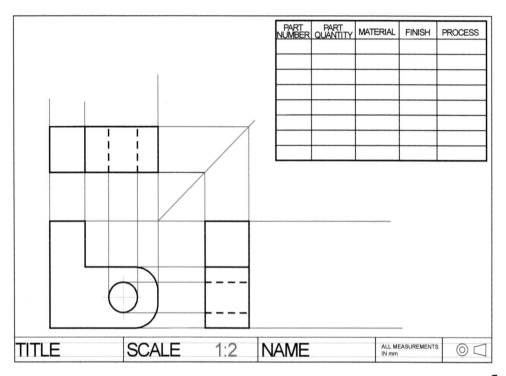

PART NUMBER	PART QUANTITY	MATERIAL	FINISH	PROCESS

TITLE	SCALE	1:2	NAME	ALL MEASUREMENTS IN mm	◎ ◁

Task 1.10

Constructing a 3rd angle orthographic projection

Things you will need to complete this task:
- 1 × Drawing kit including 45° set square and 30° set square, ruler, compass, H and B pencils
- 1 × A3 sheet of paper (landscape)

Follow the step-by-step guide on the next pages, using all the skills you have learnt so far. Try to be as accurate as possible.

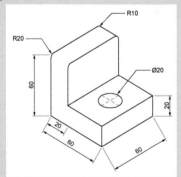

⬆ *This is the part/product you are going to be drawing. Try to keep the dimensions as accurate as possible when drawing your 3rd angle orthographic projection. All the units shown are in millimetres.*

(continued overleaf)

Task 1.10 *continued*

Step 1.	Draw a neat border, a title block, an isometric view of the object you will be constructing and don't forget the 3rd angle symbol.	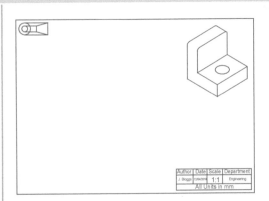
Step 2.	Making sure you have the correct overall dimensions, draw a 2D crate and **project** the lines up (for the plan view) and across (for the side view) using construction lines (**you do not have to draw in the dimensions yet**).	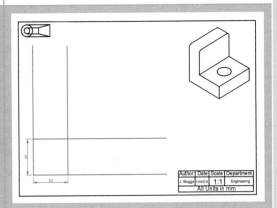
Step 3.	Within your 2D crate, draw the front view of your isometric object, making sure the dimensions are accurate.	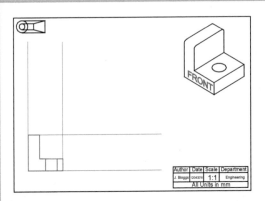
Step 4.	Where there is detail on the front view (corners, holes, etc.) **project** those details as projection lines up (plan view) and across (side view). Also, using your 45° set square, draw a 45° line from the corner of your 2D crate.	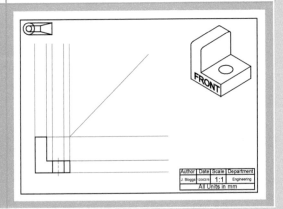

Task 1.10 *continued*

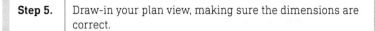

Step 5.	Draw-in your plan view, making sure the dimensions are correct.	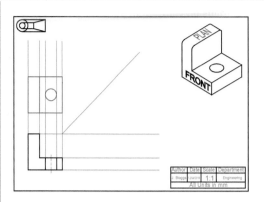
Step 6.	Where there is detail on the plan view (corners, holes, etc.) **project** those details as projection lines across to the 45° line THEN, where your projection lines intersect the 45° line, **project** the lines down to automatically create the side view.	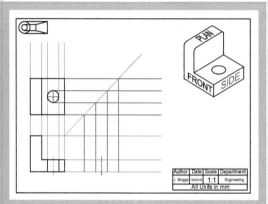
Step 7.	Complete your side view adding any further details (e.g. radius, centre lines, etc.).	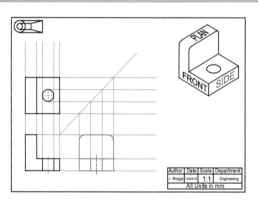
Step 8.	Use weighted lines to pick-out the three views and any other required detail (centre lines, hidden detail, etc.).	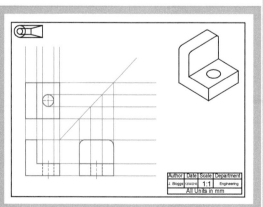

(continued overleaf)

Task 1.10 *continued*

Step 9.	Dimension your 3rd angle orthographic projection drawing using the guidelines on how to dimension accurately (see page 25). Also include a parts list, adding any details needed you think a third person would need to be able to manufacture your part/product (quantity, materials, etc.). Dimensions and the parts list can be seen in greater detail on the 3rd angle orthographic projection on page 33.	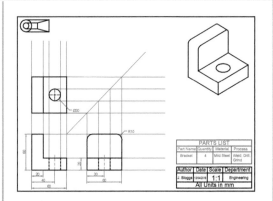
Step 10.	Add in any extra details you may think are needed (e.g. sectional view; check page 28).	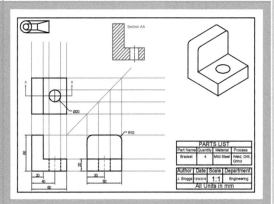
Step 11.	Check your completed drawing for any mistakes. Finished!	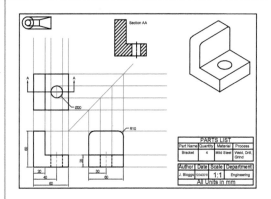

Task 1.11

In your notebook, list the items circled A–G shown on the 3rd angle orthographic drawing below.
For example: A: The drawing is in _____ angle orthographic projection.

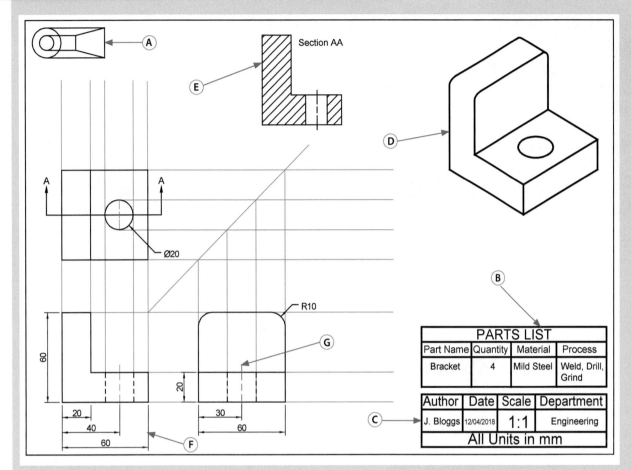

Task 1.12

Things you will need to complete this task:
- 1 × Drawing kit including 45° set square and 30° set square, ruler, compass, H and B pencils
- 1 × A3 sheet of paper (landscape)

Using your learnt knowledge, construct an accurate 3rd angle orthographic projection of the object on the right.

Success criteria

Your drawing must:
- be dimensionally accurate
- use construction and weighted lines
- be dimensioned correctly
- include a parts list
- include a title block
- include a section view
- include any other standard conventions (e.g. 3rd angle symbol, hidden detail).

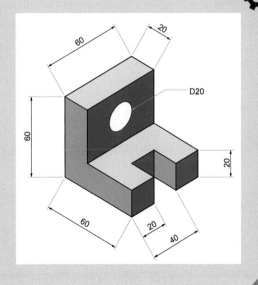

Top tip

Take your time. Focus on accuracy. If you are unsure, go back and check the lessons in this chapter.

Top tip

When you are finished and the drawing is complete, ask yourself this question: 'Can a factory manufacture this product based purely on this drawing?'

2

Communicating Design Ideas

In this chapter you are going to:

→ Learn about the need to communicate engineering information using
  - appropriate language (technical or otherwise)
  - structured methods of presentation
  - verbal and non-verbal methods
  - 2D and 3D approaches of communication.

This chapter will cover the following areas of the WJEC specification:

Unit 1 LO2 Be able to communicate design solutions	
AC2.2 Communicate design ideas	Communicate: convey meaning using appropriate language; logical structure; presentation of information; clarity of language and presentation; use of appropriate terminology; audiences (Engineers, non-Engineers); use of visual support, e.g. mock-ups, CAD

Introduction

It is an Engineer's responsibility to communicate ideas, solutions and project-related issues to different groups of people such as colleagues who could work on the project with you, clients that have given you the brief and asked for your services, or customers that will eventually purchase/use your proposed solution.

Quite often it is the client or customer that is paying for the project and they will need to understand the engineered solutions before giving their approval and allowing the project to go ahead. Customers and clients would not be able to understand a lot of the language used by Engineers due to the technical nature of the terminology used in the subject area. It is vital, therefore, that Engineers learn and use a series of communicative skills that allow them to convey meaning with clarity and transparency to different groups of people.

We can categorise these various groups of people in two ways:
- **Engineers**

and

- **non-Engineers**.

Do you think Engineers need to communicate differently from non-Engineers?
Do you think non-Engineers would understand the technical language Engineers use?

Task 2.1

Write a list of as many new 'engineering' words or phrases you have already learnt after **only one chapter**. As you complete this task, ask yourself 'would any of my non-Engineer friends or family understand these words and what they mean?'

Communicating with Engineers

Communicating with Engineers should be straightforward. As an Engineer yourself you are not restricted in the language that you use or the method in which you want to convey your thoughts and ideas. You would be free to use the technical language specific to the engineering subject area, knowing that the other party involved in the conversation (another Engineer) would fully understand your meaning.

Another advantage to communicating with Engineers would be the ability to use technical language without the need for further explanations. This would limit any errors in communication and any mistakes that may occur in the production of engineered solutions or products. Communicating with other Engineers would also be a lot faster and could reduce the amount of correspondence going back and forth between everyone involved.

However, there are accepted procedures that are commonly used by Engineers to communicate ideas and solutions effectively, which will be explored later in this chapter.

> You have already learnt some communication techniques … what are they?

Communicating with non-Engineers

Communicating accurately with non-Engineers can be a more difficult and time-consuming process. It is unlikely that non-Engineers would have been trained in the use of technical equipment and standardised engineering techniques, and therefore would not understand a lot of the vocabulary and technical terminology used.

Developing a series of techniques that allow non-Engineers to understand an Engineer's solutions and thought processes is therefore vital in ensuring the success of projects.

As students of engineering you must develop effective communication skills to ensure clarity of communication to eradicate potential costly mistakes, promote positive relationships with the customers/clients and, in turn, create successful engineering solutions.

Ways of communicating

There are many ways of communicating your thoughts and design solutions to both Engineers and non-Engineers. You can break down the areas and skills you could develop into four simple categories:

1. verbal communication
2. written communication
3. pictorial/visual communication
4. 3D shapes and models.

Verbal communication

By talking/conversing, you can describe your solution using only words. This is sometimes the simplest form of communication and is quick, easy and does not need any specialist equipment (other than possibly a phone). However, verbal communication is prone to mistakes. How many times have you misunderstood or have been misunderstood when having a conversation?

COMMUNICATION

Many Engineers have to communicate with people who speak different languages and come from different countries such as China (one of the biggest manufacturing countries in the world), where the cost of manufacture could be much lower.

Task 2.2

Find a simple product or 3D shape. With another person, Engineer or non-Engineer, use only **words**, to describe the 3D shape to the other person (do not let them see the object). The other person must now try to **draw** (paper and pencil) the 3D object as accurately as possible, using only the information they receive from your verbal description. You have **2 minutes** to complete this task.

1. How did the drawn outcome compare to the real thing?
2. What problems did you both encounter?
3. Are there alternative/better ways of communicating?
4. Are there any benefits to communicating verbally?

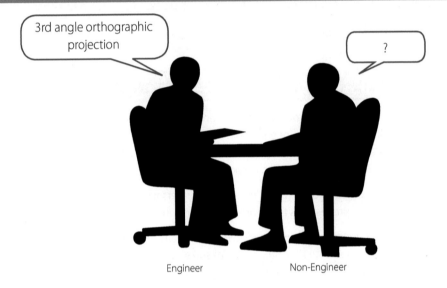

Engineer   Non-Engineer

Written communication

You could describe your solution by writing a series of annotations, sentences and paragraphs to explain and describe your engineered solutions. This can be done in the form of back-and-forth email exchanges, texting or even documents sent by traditional post.

Written communication can have some positive aspects, such as the ability to take time to craft your words and ensure mistakes are kept to a minimum. Written communication also allows clear descriptions of intentions to be laid-out in black and white and can act as a permanent record of events. Important documents are also stored for future referencing if needed.

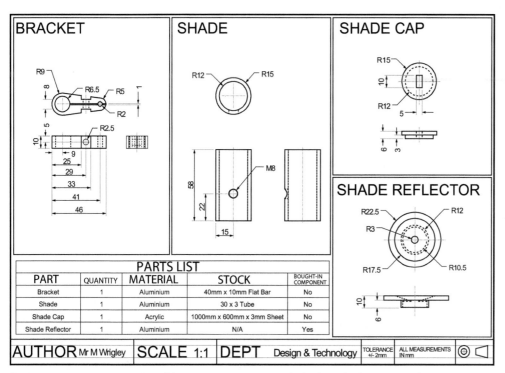

BRACKET

SHADE

SHADE CAP

SHADE REFLECTOR

PARTS LIST				
PART	QUANTITY	MATERIAL	STOCK	BOUGHT-IN COMPONENT
Bracket	1	Aluminium	40mm x 10mm Flat Bar	No
Shade	1	Aluminium	30 x 3 Tube	No
Shade Cap	1	Acrylic	1000mm x 600mm x 3mm Sheet	No
Shade Reflector	1	Aluminium	N/A	Yes

| AUTHOR Mr M Wrigley | SCALE 1:1 | DEPT Design & Technology | TOLERANCE +/- 2mm | ALL MEASUREMENTS IN mm | |

↑ *Many engineering companies still keep a paper, written record of technical information such as orthographic drawings for reference and legal purposes.*

The design brief

One of the main components of written communication in Engineering and Design (as subjects) is the design brief.

A design brief is a statement (or series of statements) stating exactly what is wanted and needed. A design brief is generally written by the customer/client and Engineer together after a series of discussions working out exactly what would be wanted and needed. A design brief is a clear 'statement of intent' that needs to be precise and written with clarity (design briefs are discussed in more detail on pages 70–72).

However, as with most forms of communication, written communication can also be misinterpreted, especially when dealing with non-Engineer customers or clients. Many descriptions might be needed to explain to clients the technical information used in modern-day engineering projects that could further confuse issues.

Can the written word be taken out of context or misunderstood?

↑ *How many times has one of your emails or texts been misinterpreted by the person you sent it to?*

Pictorial/visual communication

A common method of communication between Engineers and non-Engineers is using pictures, images and diagrams to convey meaning. Having a series of images **showing** customers and clients possible solutions is a quick and easy way to communicate effectively, as human beings process the story of an image far quicker than that of text ... after all, 'a picture paints a thousand words'.

Images can also easily show technical issues far more clearly, including full 3D objects, and even show the clients all the hidden views, mechanisms and complicated detail.

Have you ever put together some flat-packed furniture at home?
What were the instructions like when you looked at them?
Were they mostly words or pictures?

Here are some different ways of showing engineering solutions with pictures and images.

Sketches/rendered presentations

These are visual representations of design solutions and are meant to show how the solution works and how it can be developed. Quite often they have colour and lots of annotations to explain various parts of the solution. These can be done by hand or with computer aided design. They can also form part of a conversation, used to help describe and explain a project.

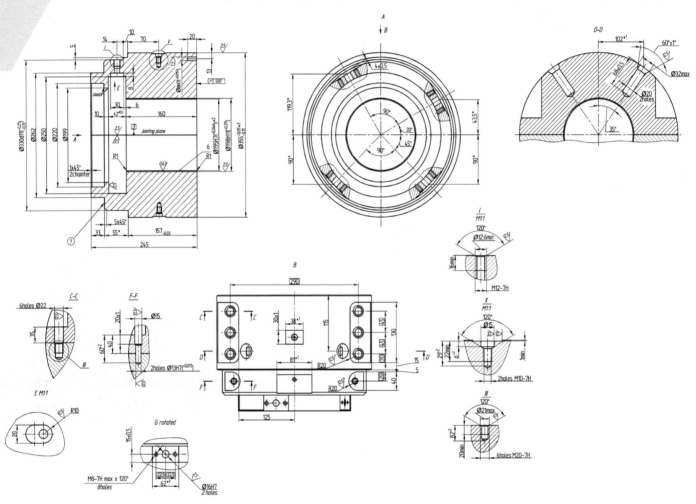

Working/engineering drawings

Working or engineering drawings, as shown on the next page, are technical drawings (orthographic projections) that show all the relevant technical details of a design solution, such as dimensions and materials. Engineers are more likely to understand this way of communicating, as they can be difficult to understand without some form of formal training.

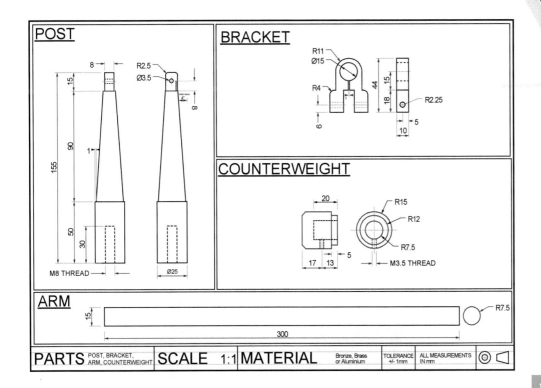

Key term

Exploded drawing: a drawing showing the parts of an object or product that have been separated to show how they interrelate or go together.

Assembly drawings

Assembly drawings are a good way of showing how multiple parts of an object or solution go together or interact with each other. This type of image is also known as an exploded drawing and is drawn in isometric view. They are commonly used as instructions for self-assembly furniture.

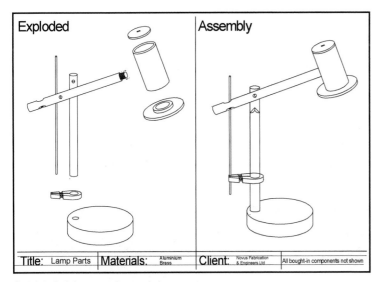

↑ *Exploded drawing of a simple lamp.*

CAD drawings/models

CAD drawings or models are created on a computer. CAD means computer aided design. This is where modern Engineers use technology (computers) to help them (aid) in the design process. CAD images can be produced in 2D or 3D, depending on what information you need to communicate and which CAD program you are using.

All modern manufacturing and engineering solutions use CAD to complete the designs before going into production.

CAD works very well when communicating with non-Engineers, as CAD can rotate digital models so they can be observed from all angles, show moving parts and mechanisms, show how the design operates under forces and also can be quickly amended to show non-Engineers the impact of any changes made.

CAD is so versatile that is it used advantageously in many professions, from taking a customer on a virtual tour of a new building, to predicting how much material would be needed for a bridge to be safe enough to cross.

As well as being a great digital communicator, CAD can also be used to create actual physical models by using CNC machines to cut-out 3D shapes from resistant materials, to the more modern approach where you can now print a 3D physical model of your digital image.

Key term

CNC machines: machines that are computer numerically controlled.

↑ *3D models created using CNC machines.*

3D shapes and models

An excellent way of communicating potential engineered solutions to non-Engineers is by using 3D shapes that the non-Engineers can pick up, hold, interact with and have a tactile experience with, in order to inform them of what the solution is. Physical models are really good for showing not only what the design solution is going to look like but also how it feels and looks in the real world. Non-Engineers can easily visualise the solution when they are interacting with a 3D model. Models, however, do not necessarily function and may not be able to show how a design actually works. Physical models can be created with any type of useful material such as paper, card, Styrofoam™ or even woods and plastics.

↑ *Models of wind turbines.*

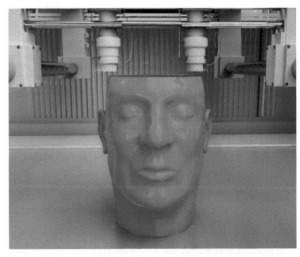

↑ *A model of a head being 3D printed.*

Prototypes

Prototypes are similar to models. However, prototypes should also demonstrate how the final solutions actually work. This includes electronics and moving parts. Prototypes are the most expensive and time-consuming way of communicating ideas but are also the closest to the real thing. Prototypes are a very good idea for those products that will cost a lot of money to set up and start making, to ensure that everything is absolutely correct before beginning.

↑ *A prototype car.*

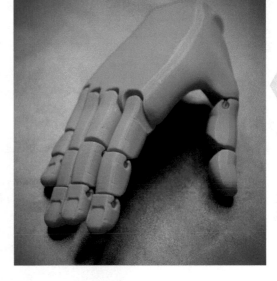

→ *A prototype prosthetic hand that has been 3D printed.*

Task 2.3

Now you understand the different ways of communicating technical information effectively to non-Engineers, try answering these questions:
1. Which way of communicating is best?
2. Can/should you use only one option?
3. Can you mix them up by using a combination of communication models/procedures?
4. Who would benefit best from each model?
5. Is communication dependent on what skills you have learnt?

Try producing some ideas for a new 'communication device' (it does not have to be a phone), using sketches and annotations. Use the annotations to describe what you are trying to show in your sketches.

AND

Either by hand or by using a CAD software program, create a presentation drawing of a product. Try to communicate as much information as you can about the product (annotations, etc.). It can either be in 2D or 3D.

AND/OR

Try creating a simple model of your solution. You can use card/paper/modelling foam or even CAD if you have access to the software.

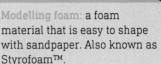

Key term

Modelling foam: a foam material that is easy to shape with sandpaper. Also known as Styrofoam™.

3 Materials and Properties

In this chapter you are going to:
→ Learn about PROPERTIES of materials
→ Understand the different CATEGORIES of materials
→ Identify materials of named products
→ Explain WHY materials were chosen for specific tasks
→ Understand and name specific SMART materials
→ Understand about material FINISHES (categories and examples).

This chapter will cover the following areas of the WJEC specification:

Unit 1 LO1 Know how engineering products meet requirements	
AC1.1 Identify features that contribute to the primary function of engineered products' features	Properties of component materials
Unit 2 LO2 Be able to plan engineering production	
AC2.1 Identify resources required	Resources: materials; equipment; tools; time
Unit 3 LO1 Understand effects of engineering achievements	
AC1.1 Describe engineering developments	Materials
Unit 3 LO2 Understand properties of engineering materials	
AC2.1 Describe properties required of materials for engineering products	Engineering products: structural, e.g. buildings, bridges; mechanical, e.g. gearbox, crane, bicycle; electronic, e.g. mobile phone, communications, alarm Properties: tensile strength; hardness; toughness; malleability; ductility; conductivity; corrosive resistance; environmental degradation; elasticity
AC2.2 Explain how materials are tested for properties	Tests: destructive tests; non-destructive tests
AC2.3 Select materials for a purpose	Materials: ferrous; non-ferrous; thermoplastics; thermosetting plastics; smart; composite

Introduction

In this chapter you are going to find out all about different materials and what some of those materials can do. Materials are the fundamental make-up of the world we live in, so as an Engineer it is your job to discover what the capabilities are for each of the materials you intend to use.

When we talk about **MATERIALS** we are really talking about **PROPERTIES**.

So what are material properties?

The properties of a material explain to us exactly what the material does. In other words, what that material is good at doing and what it is bad at doing. Once we understand what materials can or cannot do then we can begin selecting the right ones to perform different tasks and understand HOW and WHERE they can be used.

For example, when asked to create a structure that allows traffic to cross a river or gorge, an Engineer would firstly discover what properties would be needed in order to be strong and rigid enough to span the gap and support the traffic. Once those properties were identified then they would pick the appropriate material.

The ancient Roman Engineers were masters of understanding materials and properties. By using their knowledge effectively, they were able to quickly build bridges that allowed them to march their armies all over Europe, conquering the known world. This gave the Romans a huge advantage, as, without this knowledge, other cultures were severely restricted in their movements.

Properties of materials

The property of a material dictates how it will perform and react to the environment it is functioning in and how it will react to the job you have asked it to do.

Steel bridge structures = good

Chocolate kettle = bad

Following are some properties and their definitions. This list uses specific names/words to identify a property. Learn these words and their meanings, as Engineers have to describe material properties all the time for each project they undertake.

Elasticity	Ability to regain its original shape (e.g. rubber).
Ductility	Ability to be stretched without breaking (e.g. copper).
Malleability	Ability to be pressed, spread-out or hammered (e.g. lead).
Hardness	Ability to resist scratching, cutting or wear-and-tear (e.g. high-carbon steel).
Work hardening	Property changes due to working (e.g. bending steel back and forth).
Brittleness	Will snap easily and will not bend (e.g. glass).
Toughness	Resistant to breaking and bending (e.g. cast iron or urea formaldehyde polymer).
Tensile strength	Retains strength when stretched (e.g. some aluminium alloys).
Compressive strength	Very strong under pressure (e.g. concrete).
Corrosive resistance	If it will corrode in the environment it is working in (e.g. iron rusts).
Conductivity (electrical)	Ability to conduct (transmit) electrical current (e.g. copper wires).
Conductivity (thermal)	Ability to conduct heat (most metals such as steel cooking pans).
Environmental degradation	How the material corrodes and degrades in an environment (salt water, weather, fire).

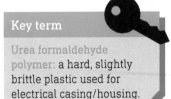

Key term

Urea formaldehyde polymer: a hard, slightly brittle plastic used for electrical casing/housing.

Task 3.1

Below are some statements relating to material properties. Copy them into your notebook and have a go at filling in the blanks.

1. We use cast iron for manhole covers because it is _____ and therefore is unlikely to bend or break when traffic drives over it.
2. We have some very nice copper-bottomed pots and pans at home for cooking in and they are great at _____ the heat from the stove.
3. Don't throw stones near glass windows. They are _____ and likely to break.
4. My drill-bits are made from _____, which makes them very _____ and good for cutting through other metals.
5. Most modern bridges are made from _____ and _____, as they have great _____ and _____ properties and can therefore bear the weight of all the traffic.

Materials

Now we understand about the properties of materials we can begin to identify the materials themselves, what categories they fall under and what functions they can perform (depending on the properties they have).

In this section we are going to look at:

- metals
- plastics
- composite materials
- smart materials.

Metals

A large range of metals exist in the world, all with different, useful properties and all used to perform different tasks, from the gold and copper used to produce your mobile phones to the steel and aluminium used to create the skyscrapers seen in the world's largest cities. Metals generally fit into **two** categories:

- **ferrous** and
- **non-ferrous**

and the sub-category:

- **alloys** (that are made from the ferrous and non-ferrous metals).

Ferrous metals – metals that contain Iron

The word FERROUS comes from the Latin word *ferrum*, which literally means iron.

Therefore, ferrous metals contain iron. Rather than just being 100% iron they tend to be made-up from iron (e.g. STEEL can be 99.9% iron and 0.1% carbon).

Iron is the fourth most common element (second most common metal) in the world's crust and is therefore easy to find. Pure iron tends to be too soft to use on its own, so other metals and elements are mixed with it to create useful materials (e.g. steel).

You can easily identify a ferrous metal in one of two ways. Firstly, iron corrodes (rusts), so anything with rust on the surface (oxidisation) must have iron in it. Secondly, iron has magnetic properties and can easily be identified if you have a magnet to hand. *However, some metals that contain iron have corrosive-resistant properties and may not rust (e.g. stainless steel).*

↑ *A steel-making foundry.*

Latin, *ferrum* = iron
English, ferrous = contains iron

Key term

Oxidisation: the process of oxidisation; where steel/iron surfaces react with the atmosphere and create ferric oxides (rust).

Examples of FERROUS METALS include:

	Material	Properties	Common uses	Made-up from
	Mild steel	• Good tensile strength • Good toughness • Corrodes easily	Used in many products such as: • carcasses for PCs • Xboxes, etc.	• Iron • 0.1–0.3% carbon
	High-carbon steel	• Tough • Hard • Can be brittle	Tools such as: • saw blades • drill bits	• Iron • 0.5–1.5% carbon
	Stainless steel	• Corrosive resistant • Tough	• Medical instruments • Cutlery	• Iron • Nickel • Chromium
	Cast iron	• Good compressive strength	• Drain and manhole covers • Engine blocks	• Iron • 2–6% carbon

Non-ferrous metals – metals that do not contain iron

Non-ferrous metals do not contain any iron. There are lots of examples of non-ferrous metals (including various ALLOYS) such as aluminium, gold and copper.

Non-ferrous metals can have different PROPERTIES from iron and have many different uses, copper being a great conductor of heat and electricity (electrical cables/wires). Non-ferrous metals also tend to have much greater resistance to corrosion than ferrous metals and do not have magnetic properties.

However, most non-ferrous metals are not as common as iron (with the exception of aluminium, which is the first most common metal in the Earth's crust) and all non-ferrous metals tend to be much more expensive to refine from the metal ore.

Non-ferrous metals are also more expensive to fabricate compared to iron.

> **Key term**
>
> Fabricate: to shape and join materials to create a product.

For example, a rotary washing line is an engineered product that performs a task in an external environment. It is likely the washing line is going to get rained on and would therefore probably encounter environmental degradation. You can purchase rotary washing lines that are made from aluminium, as they are corrosion resistant and light (non-ferrous). However, you can also purchase steel (99.9% iron and 0.01% carbon) washing lines that would rust and are very heavy. So why purchase a steel washing line? Steel is a much cheaper material and is also much easier to fabricate. Steel can easily be welded with common welding equipment, whereas aluminium would need specialist welding equipment. All these factors go towards the product being cheaper to purchase for the consumer.

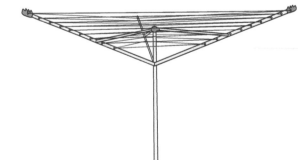

Examples of NON-FERROUS METALS include:

	Material	Properties	Common uses	Made-up from
	Aluminium	• Light • Soft • Malleable	• Good for alloys • External products • Aircraft	• Aluminium
	Lead	• Ductile • Malleable • Heavy	• Roofing • Batteries	• Lead
	Copper	• Good conductor • Ductile	• Piping • Electrical wiring	• Copper
	Gold	• Soft • Malleable • Tarnish/corrosion resistant	• Jewellery • High-end stereo connections	• Gold
	Brass	• Hard • Corrosive resistant	• Musical instruments • Ornamental products	• Copper • Zinc

Alloys

↑ *Examples of Bronze Age axe-heads.*

An ALLOY is a mixture of elements that usually have a metal as the major/parent component (e.g. STEEL is 99.9% IRON and 0.1% CARBON). Alloys were developed to create different properties from just having a pure, parent metal. By heating-up, melting and mixing different metals you can create new metals with new, different properties.

Bronze is an alloy that is created by mixing copper and tin. Bronze is harder, more corrosion resistant and easier to melt and cast into different shapes (e.g. axe-heads) than both its parent metals, copper or tin.

Duralumin is a modern alloy created from having aluminium as a parent metal then adding small amounts of other metals (copper, magnesium, manganese) to create a material that is lightweight, strong and extremely corrosion resistant.

→ *Many car and aircraft parts are made from duralumin, an alloy.*

Examples of ALLOYS include:

	Alloy	Made-up from	Common uses
	Duralumin	• Aluminium • Copper • Magnesium • Manganese	• Car parts • Aircraft parts
	Brass	• Copper • Zinc	• Musical instruments • Ornamental products
	Stainless steel	• Iron • Nickel • Chromium	• Medical instruments • Cutlery

Alloying agents/elements

Modern Engineers use many different alloys that perform all different types of tasks. Different **alloying elements** have also been discovered and what can be created when they are mixed with metals. We also know what ratios of metals and elements are needed to be mixed to create specific properties.

There are some common alloying elements that are used in many modern-day practices.

Following are some examples and what properties they can add to an alloy:

Alloying element/agent	Properties
Nickel	• Increases strength • Increases hardness • Increased resistance to corrosion
Chromium	• Increases hardness • Increased resistance to corrosion • Increased toughness
Vanadium	• Increased toughness of steel • Increased wear resistance

↑ *Engines are often made from chromium.*

Supply of metals

When Engineers use metals, they need to know what forms they can be supplied in to ensure that the correct shape is ordered for the engineering project that is being undertaken, as well as to make sure the correct terminology is used when ordering the materials.

The various metal shapes you can purchase are mostly going to be **extrusions**. An extrusion is a fixed sectional shape (think back to sectional drawings) that is continued for a desired length (imagine pushing Play-Doh or Plasticine through a fixed shape).

Extrusions are very common shapes used in the world; how many can you see from where you are sitting?

The diagrams below show examples of metals and how they are supplied. The dimensions of the cross-section can vary, whereas they all tend to be extruded to desired lengths.

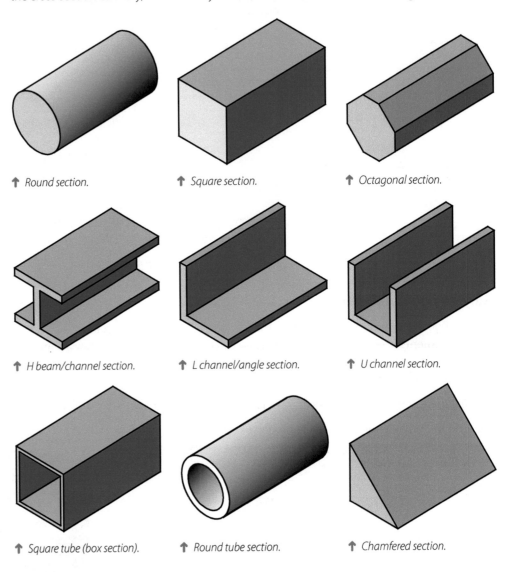

↑ *Round section.* ↑ *Square section.* ↑ *Octagonal section.*

↑ *H beam/channel section.* ↑ *L channel/angle section.* ↑ *U channel section.*

↑ *Square tube (box section).* ↑ *Round tube section.* ↑ *Chamfered section.*

Plastics

Imagine a magical material that you could mould into any shape you can imagine, make any colour you want, that would never rust, is very low cost, that you could apply any finish to (rough, gloss) that you can think of and is also very strong and very lightweight.

Welcome to plastics.

Regardless of all the negative impacts plastic has on the world, the reason it is used so widely is because of all the amazing properties it has. It is easy to create engineered solutions to problems when you have a material that can do all these incredible things. This is why so many Engineers and Designers specify its use.

So where do plastics come from?

Plastics are made from the chemicals that are extracted from crude oil. Crude oil is extracted from the ground and transported to a refinery where it goes through the process of being refined. Many different chemicals are extracted from the refinery process, one of which is naphtha. Naphtha then is processed further to produce plastics using the polymerisation process.

Key term

Polymerisation: the industrial process used to create plastics from naphtha.

Following is a diagram showing the relatively simple process of how naphtha is extracted from crude oil.

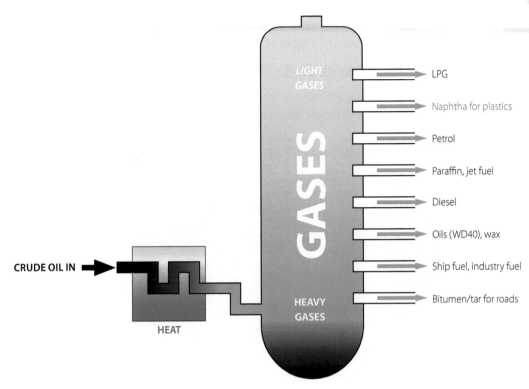

⬆ *A simple example of how crude oil is refined.*

From naphtha we then derive monomers, which are then joined using the polymerisation process to create polymers.

Plastics can be separated into two main categories:
- **thermoplastics**
- **thermosetting plastics**.

Thermoplastics

These plastics can be re-shaped when re-heated and are therefore re-mouldable. This can be done multiple times before the thermoplastic starts to degrade. Thermoplastics are also recyclable.

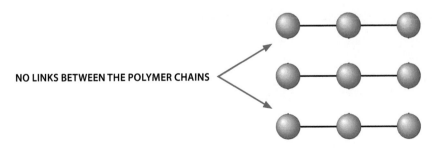

↑ *A hard-wearing protective cover made from acrylic, for a display.*

Examples of THERMOPLASTICS include:

ACRYLIC

Also known as Perspex®. Quite hard-wearing and used for many applications such as signs, glass substitute phone covers and kitchens.

HIGH-IMPACT POLYSTYRENE (HIPS)

A tough, rigid plastic that is often used in the vacuum-forming process. Good for packing (biscuit tin inserts), toys and cutlery dividers/draw organisers.

↑ *A cutlery divider made from HIPS.*

PVC

PVC is a hard-wearing plastic used for doors and windows (UPVC), waste pipes and electrical tape. It is also used for many other products such as plumbing fittings, electrical wiring and products used in the medical industry.

↑ *UPVC windows are hard-wearing and can be UV (light) resistant.*

NYLON

Nylon has excellent low-friction properties. Great for assemblies with moving parts, especially for products such as door runners, cogs/gears and washers.

⬆ *A gear made from nylon will have natural lubricating properties without the need to oil and grease it.*

Thermosetting plastics

Unlike **thermoplastics**, **thermosetting** plastics are joined across the polymer chains. This gives **thermosetting** plastics a very strong bond between the **monomers**.

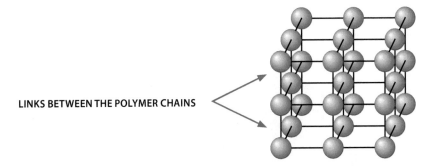

LINKS BETWEEN THE POLYMER CHAINS

Thermosetting plastic cannot be re-heated and re-moulded like thermoplastics.

Examples of THERMOSETTING PLASTICS include:
* epoxy resin
* urea formaldehyde
* melamine formaldehyde.

EPOXY RESIN

Often used as a type of glue, epoxy resin is good for laminating (layering) materials to create products such as skateboards. Also known as the glue Araldite®.

⬆ *Epoxy resin is often used as a binding agent and surface cover for snowboards and skateboards. It starts as a liquid and hardens into a solid.*

UREA FORMALDEHYDE

A hard, slightly brittle plastic used for electrical casing/housing such as plug sockets and smoke alarms.

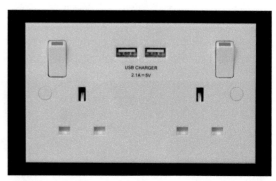

⬆ *Many electrical components are made from urea formaldehyde as they will not deform (change shape) when under high temperatures. Great for electrical safety.*

MELAMINE FORMALDEHYDE

A hard plastic with good heat-resistant properties. Used for microwavable dishes and other products that are exposed to heat.

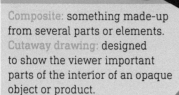

⬆ *Often, melamine formaldehyde is used for children's plates and dishes, as it is very tough (toughness) and they are sometimes marketed as 'shatter-proof'.*

Composite materials

Composite materials are materials made from two or more materials in order to add extra properties (e.g. make it stronger). Unlike alloys, composite materials are not combined after melting or mixing but are kept separate and usually bonded together with glues, adhesives or resins. This process can add extra properties where needed in the new composite material.

Below is a cutaway drawing of a composite front door. Many homes now have these types of composite products/materials as their exterior doors and the annotations explain why.

> **Key terms**
>
> Composite: something made-up from several parts or elements.
> Cutaway drawing: designed to show the viewer important parts of the interior of an opaque object or product.

Polyurethane high-density expanded foam core:
* extremely lightweight
* sound proofing
* fire resistant
* insulating
* low cost

Hardwood timber frame:
* good tensile strength
* strong and rigid (creating a good, strong framework)
* good strength-to-weight ratio

Fibreglass (GRP) outer sheet:
* fibreglass (composite) made from glass fibres and epoxy resin
* good strength-to-weight ratio
* self-finishing
* weather/corrosion/UV resistant
* mouldable into any shape
* colour/texture choices

> **Key term**
>
> GRP: glass reinforced plastic, also called fibreglass. A mix of fibreglass and epoxy resin.

There are many types of composite material that are commonly used in everyday products. Examples of COMPOSITE MATERIALS include:
* man-made/manufactured boards
* GRP
* carbon fibre.

MAN-MADE/MANUFACTURED BOARDS

Made from timber off-cuts, chips, fibres and glues/adhesives. Manufactured boards include plywood, chipboard, MDF and blockboard. They are **strong, resistant to warping, low cost** and can come in large surface area sheets/boards.

GRP

GRP is made from glass fibres and epoxy resin. Used in many products such as boat hulls, waterslides and telecoms street cabinets (big green units). It has a good **strength-to-weight ratio**, can be **moulded** to almost any shape and is **UV resistant**.

↑ *A sample of MDF.*

CARBON FIBRE

Made from carbon fibres and resins, it is used in many products such as sports equipment, motorsports (F1), the aerospace industry and safety equipment. It is extremely **lightweight**, has a fantastic **strength-to-weight ratio** and can be **moulded** to almost any shape.

↑ *Resin being sprayed onto the fibreglass.*

↑ *Exhaust pipes and bumper made from carbon fibre.*

Smart materials

Smart materials are materials that change their properties and characteristics according to changes in the environment (e.g. temperature, light, force, etc.) they function in. Smart materials are being used more often in products and processes as they can complete multiple tasks due to their changing nature.

Have your ever seen a mug change colour when hot water is poured in?

MUG

HOT TEA

MUG WITH A CHANGED ENVIRONMENT

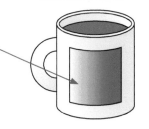

Change in properties

The environment has changed by adding HEAT. The SMART MATERIAL used on the mug is known as thermochromic ink, meaning ink that changes colour according to the temperature.

What do you think HYDROCHROMIC and PHOTOCHROMIC materials do?

Examples of SMART MATERIALS include:

* nitinol
* D3o
* polymorph.

NITINOL

Nitinol is an SMA that can remember a shape when heated. It is an alloy of nickel and titanium and is used for stents (holding open arteries), under-wire bras and glasses.

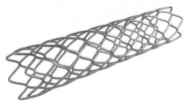

↑ A 'stent' made from nitinol, an SMA. It is used for keeping arteries open and can be inserted into an artery when small and compact, and then expands and gets bigger when inside the artery (when the environment changes).

↑ Some glasses spring back to their original shape if sat or stepped on. They are made from an SMA.

D3o

A lightweight, soft flexible and malleable material that stiffens when subjected to sudden force. Used for sports clothing, including skaters' beanies (headwear) and motorbike riders 'leathers'.

↑ A biker's jacket with D3o panels.

POLYMORPH

A hard polymer (plastic) which softens and becomes malleable at 62° Celsius (°C). Used for tool handles and individual grips.

↑ Polymorph granules.

↑ A mask made from polymorph.

> **Key term**
>
> SMA: shape memory alloy. A metallic alloy with a 'memory'.

Testing of materials and their properties

When the process of selecting materials is complete, Engineers begin the testing process to see if the properties the materials have are sufficient to complete the task asked of them. Sometimes the completed solution/product is tested in its functioning environment.

It is very important that Engineers test materials for the correct properties, as selecting the wrong materials or inadequate properties could result in costly mistakes. Imagine a bridge collapsing because the materials were not tested properly.

Materials are also tested to define the limits of their operating parameters (e.g. how much load something can take before it starts to break). Once those parameters are understood, Engineers can start designing into their product a factor of safety (FoS). For example, when the failure point of a product/material is found, Engineers can add more material to take double the load, essentially increasing the FoS. Testing the limits of a material and product allows Engineers to ensure their engineered solutions are very safe to use.

By going through the testing process Engineers can:
* save money
* meet health and safety needs/standards
* prevent product failures
* provide data for future projects/innovations.

Testing for materials properties can fall under two categories:
* **Destructive tests**: where the materials (or products) are put under forces until they begin to fail and catastrophically fail. These tests can check for properties such as tensile strength, compressive strength, toughness, hardness, etc.
* **Non-destructive tests**: where the materials (or products) are tested without damaging the materials or products themselves. Non-destructive testing allows you to test your material/product in its functioning environment to see how it performs in its day-to-day life. Non-destructive testing can save money by not having to prepare and destroy materials and products. This testing process also allows for testing the integrity of historical structures.

CAD and testing

Most good computer aided design programs now come with the ability to *virtually* test materials and products in virtual environments. Modern CAD programs already contain data for materials and you can specify what material your CAD models are made from. If your product is going to fail under a load then it can show in colours (mainly red) where your product would fail. This allows Engineers to redesign their solutions before going into production. This process is called FEA.

Finishes

Quite often, the materials that have been used need to be **finished** as part of the processes used to create them. This means that the materials need to undergo another manufacturing process to give them a finish. Reasons for finishing a material can differ:
* aesthetic (make it look nice)
* functional (e.g. give it a rougher surface for grip)
* protection/corrosion resistance (stop it rusting or discolouring).

Examples of finishes for metals

Plastic dip coating: used mainly on steel. Metal is heated and dipped into plastic powder that melts and sticks onto the surface of the metal. Good for anti-corrosion. A range of colours is available for aesthetic purposes.

Key term

FoS: factor of safety. To build in a safety margin when designing products.

Key terms

Force: the push or pull on an object causing it to change velocity (to accelerate).
Catastrophically fail: to be tested until it is broken and does not work any more.

Key term

FEA: finite element analysis. A way to test your material (element) choice under forces on a computer model.

↑ *FEA of a design bike frame. Red areas are where the frame is likely to fail (under a specified load).*

Key term

Aesthetic: the way something looks to the eye.

↑ *A plastic dip-coated screw.*

Galvanising: coating a ferrous metal (steel) with zinc (non-ferrous) to protect it against corrosion. Galvanising creates a thin, non-ferrous layer of metal between the steel and the elements (rain/wind). It can be used for streetlights and fences as it has a very durable finish.

Anodising: aluminium is placed in a bath of acid where an electric current is passed through it. Coloured dye is then added, which permeates the surface of the aluminium, adding colour for aesthetic purposes as well as adding to the corrosion resistance of the aluminium part.

⬆ *Anodised aluminium carabiners.*

⬆ *A galvanised steel fence.*

Powder coating: similar to plastic dip coating but the plastic powder is sprayed on and is used more in Industry. This process is mainly used for white goods such as washing machines, dish washers, etc.

Blueing: steel is heated up and then dipped in oil. The oil permeates the surface of the steel to create an anti-corrosion layer that protects the steel against rust. The finish tends to be a blue/black colour. 'Cold blueing' can also be completed by using chemicals instead of heat and oil. Used for tools and by gunsmiths.

Painting: paint creates a barrier between the metal surface and the elements for corrosion resistance, as well as having lots of colour options for aesthetic purposes. Metal would need to be prepared first using a 'primer'. Some paints are designed specifically for metals, for example Hammerite. This finish would need to be continually maintained on products such as bridges, ships, goalposts, etc.

⬆ *Powder coated panels on a white washing machine.*

Key term

Corrosion: oxidisation of a metallic surface or rust.

⬆ *A painted steel ship.*

Enamelling: high temperatures are used to melt powdered glass onto a metallic surface to create a glass barrier between the metal and surface for corrosion resistance and aesthetic appeal. Lots of colour options are available. One of the most common items that is enamelled is the tin mug, but a lot of jewellery is also created using enamelling.

← *Enamelled mugs.*

Examples of finishes for wood

Painting: one of the most common finishes for wood surfaces. The wood needs to be prepared first with a primer or undercoat before applying a finishing coat of paint. The finish resists weather wear and many colour options are available for aesthetic appeal. Paint can be water based or oil based, and brushed or sprayed on.

Varnish: tends to be applied when the 'grain' of the wood wants to be seen (instead of covering it up with paint). Most varnishes tend to be transparent so you can still see the wood and have a gloss or matte finish. Multiple layers can be applied that create a barrier between the elements and the wood itself. Polyurethane or 'yacht' varnish is very hardwearing and is used in the marine industry.

↑ *Painting wood.*

↑ *Varnishing wood.*

Stains: wood stains tend to have a more 'watery' consistency and are generally applied with a brush. Stains permeate the surface of the wood and protect against the elements, along with adding different colour choices for aesthetic purposes.

Wax: applying wax to wood creates a waterproof finish to the surface, which can be buffed/polished to create a very smooth, natural-looking finish. Wax is good for indoor products and is aesthetically pleasing if you want to see the natural tones of the wood itself.

↑ *Staining wood.*

↑ *Waxing wood.*

Finishes for plastics

Plastic is a material with lots of great properties (strength-to-weight ratio, mouldable into any shape, various colours) and it also has the wonderful property of being **self-finishing**. Unlike the other materials discussed, plastic, with its ability to self-finish, means it does not have to go through another process to finish it. For many companies this means saving a lot of money as they would not have to purchase extra equipment, invest in extra space, or train and pay more staff to finish their plastic products.

The 'finish' on plastic products is chosen during the design stage and the interior on the plastic 'mould' would have that chosen finish on it. So, when the plastic product comes out of the mould it also retains the surface finish of the mould.

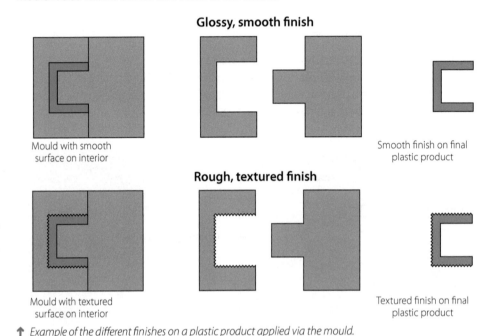

Glossy, smooth finish

Mould with smooth surface on interior

Smooth finish on final plastic product

Rough, textured finish

Mould with textured surface on interior

Textured finish on final plastic product

↑ *Example of the different finishes on a plastic product applied via the mould.*

However, when working with some production processes, plastic products may need to be finished. Examples include:

- **Cutting or sawing plastic**: this process could leave a rough edge with plastic burrs and may need smoothing-out with fine-grade abrasive paper or even buffing on a polishing machine.
- **3D printed products**: these plastic products tend to be made from plastic 'wire' that leaves ridges all around your product. This can be finished with fine abrasive paper.

↑ *Cutting plastic with this CNC machine could leave rough edges.*

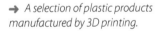

→ *A selection of plastic products manufactured by 3D printing.*

Task 3.2

Copy the following tables into your notebook and use the knowledge you have gained about materials and properties to complete them.

Ferrous metals		
Name	Product it is used for	Properties needed

Non-ferrous metals		
Name	Product it is used for	Properties needed

Alloys			
Name	Parent metal	Product it is used for	Properties needed

Polymers (plastics)			
Name	Thermo or thermosetting	Product it is used for	Properties needed

Smart materials		
Name	Product where it is used	Properties needed

(continued overleaf)

Task 3.2 *continued*

Composite materials		
Name	Product it is used for	Properties needed

Finishes		
Product	Possible finish	Why use this finish?
Goal posts		
Streetlights		
Brake lever for mountain bike		
Hook for garden shed		
The wooden deck of a sailing boat		

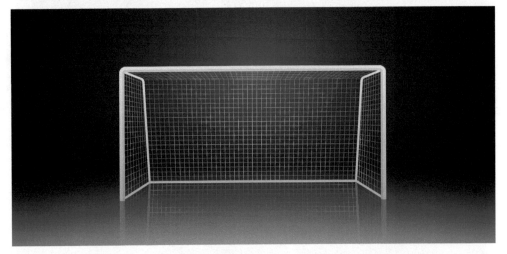

→ *What finish would you use for this goalpost (above) and boat deck (below)?*

Identifying Features of Working Products

In this chapter you are going to:
→ Learn how to analyse existing products effectively
→ Learn how key words can be used to identify and analyse products
→ Understand the term 'reverse engineering' and how to do it
→ Learn the differences between the component parts of a product
→ Understand how different parts of a product work together.

This chapter will cover the following areas of the WJEC specification:

Unit 1 LO1 Know how engineering products meet requirements	
AC1.1 Identify features that contribute to the primary function of engineered products' features	Features: of component parts; electrical components; mechanical components; properties of component materials
AC1.2 Identify features of engineered products that meet requirements of a brief	Requirements: aesthetic; environment (where used); user/customer/client; cost; safety; ergonomics; size; limits; sustainability
AC1.3 Describe how engineering products function	Function: how components interrelate
Unit 1 LO3 Be able to propose design solutions	
AC3.1 Develop creative ideas for engineered products	Creative ideas: identify features of other engineered products

Introduction

Engineers tend to be creative and innovative problem solvers. They use their skills and knowledge to specify materials and develop designs, and are key workers when creating new products. However, Engineers also develop skills that enable them to learn from past and existing products. By learning from other solutions, Engineers can improve their knowledge base and skills that will, in turn, enable them to optimise their engineering solutions and create better, more efficient products in the future. This process of 'finding things out' is also known as research.

There are several ways of learning from/ researching existing products and here you will learn about two of the most common techniques:
* product analysis
* reverse engineering.

Key term

Research: the process of finding things out.

← *Research helps you to learn from others' experiences.*

Product analysis

Product analysis is a way of analysing products against specific criteria.

So, you could **look at** and **feel** an existing product (e.g. the mobile phone in your pocket) and then ask: 'Does it look good? Is it easy to hold? Is it easy to use? Is it heavy or light?' And by asking these questions you begin to find answers that would help you create a mobile phone with all the best features from the one you have analysed.

What other questions should you ask when analysing existing products?

Fortunately, there exists a simple model you can use that contains many of the headings and titles that could be used to analyse a product. However, be aware that this is a simple model and only has the most basic titles. Adding more titles (asking more questions) is also good practice as it allows you to gather even more detailed information.

To analyse a product a good starting point would be to use the model ACCESSFM, as shown in the following table:

> **Key term**
>
> Criteria: specific headings or titles.

> **Key term**
>
> Target market: the group of users you will be designing for.

ACCESSFM		
A = AESTHETIC	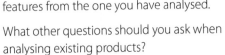	• The way the product looks. • Does the target market think it looks good? • Has it been finished well with appropriate colours?
C = COST		• How much does the product cost to buy? • How much to make? • How much to run? • Can the target market afford it? • Is it a premium or low-cost product?
C = CUSTOMER		• Who is the customer/target market? • What appeals to the customer/target market and why? • Do the customers/target market need or want the product?

(continued overleaf)

ACCESSFM *continued*

E = ENVIRONMENT		• Where will it be used? • How will it be used there? • What resources were used to produce it? • Is it a sustainable product? • Can it be recycled? • Does it have a negative or positive impact on the local and larger environment?
S = SAFETY		• Is the product safe to use for the customer/target market or in its environment? • Does it meet all safety regulations? • Is it safe to make?
S = SIZE		• What numbers or values can you attribute to the product? • Why does it need to be that big/small? • What sizes would be applicable to the customer/target market? • Can the product be scaled (up or down) to make it better?
F = FUNCTION		• Does the product function/work well? • Does it perform the job it was meant to? • Does it work for the customer/target market? • Does it function well in its environment?
M = MATERIALS AND MANUFACTURING TECHNIQUES		• What materials have been chosen to make the product and why? • What properties are needed from the materials to allow the product to function well? • Are the materials renewable/non-renewable? • Are they recyclable? • What processes were used to make the product? • Are the processes suitable or can you specify better suited ways of making the product?

Key term

Sustainable: can the manufacture of the product be maintained (is it made from renewable resources)?

Top tip

As mentioned, ACCESSFM is a good starting model to use when asking questions. However, for greater detail you may want to analyse a product using additional titles.

Maybe ACCESSFM does not have the title you need to find something out, so you may need to add your own specific titles and headings when analysing a product.

Here are some other headings and titles you may wish to consider:

Further headings		
ANTHROPOMETRICS (Greek: *ANDROS* = man, *METRON* = measurements)		• What size is your target market and how would that affect the sizes and positions of your product (think of a child's car seat)? • There are lots of anthropometric data you can access on the internet that would be useful if you are creating furniture for children.
ERGONOMICS (similar to FUNCTION in ACCESSFM)		• Does your product function well in relation to the customer/target market? • Will the sizes, positions, textures, finishes, material choice, be affected by who you are designing for (target market)? • Does the product function in its environment? Is it easy/comfortable to use?
USER REQUIREMENTS		• Does the product meet all the needs of the potential users? • Is there anything the product does not do well for the users?
POWER/ENERGY USE		• How is the product powered? • Is it powered by battery, mains or solar? • Does it need a lead? • Is it portable power? • Can it be re-charged?
PRODUCTION LEVELS		• How was the product produced? • Is it quick and one of millions in a factory or did a craftsperson make it over a longer period? • How will this affect cost, etc.?
LEGAL REQUIREMENTS	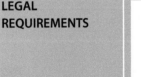	• Is this product a copy or being copied? • Are there any innovations you can or cannot use?
LIMITATIONS/ CONSTRAINTS		• What constrains the product? • Are there limits to its function? • Is the product limited to an environment/ customer/time to use?

Key term

Innovation: from the Latin *innovare*, meaning 'to make new'. Take something that already exists and improve it.

Reverse engineering

Reverse engineering is a very similar process to product analysis. With reverse engineering you are still analysing products to gain information that could be used to improve your engineering solutions. You can also use headings and titles to analyse the product. However, instead of just product analysis, reverse engineering dismantles a product piece-by-piece to discover things such as how it was put together, how it was manufactured, how all the component parts interrelate, where and what component parts are hidden and where are the hidden, innovative parts that could be vital for the product to function. Quite often, Engineers approach the reverse engineering task from two different perspectives:

- **external analysis**

and

- **internal analysis**.

The external analysis of a product can focus on (but is not be restricted to) the information the product gives to the user such as look, feel, interactive areas (buttons/levers, etc.), properties and materials.

The internal analysis of a product can focus on (but is not restricted to) the manufacturing processes, assembly, hidden components, fixings, maintenance, materials, properties and how the components interrelate with each other.

Following are some headings and titles you can use to analyse an existing product when using reverse engineering:

External:
- ACCESSFM (plus others).

Internal:
- ACCESSFM (plus others)
- components
- assembly
- repair/maintenance
- recycling
- interrelation
- innovation.

Key term

Product analysis: looking, feeling and maybe using the product to see how it works.

↑ *Recycling is an important consideration when analysing products.*

Task 4.1

Using your knowledge of product analysis and reverse engineering complete the following task.

Look at the images of vacuum cleaners below and answer the following questions:
1. Which one is the Dyson?
2. Why are they all so similar?
3. Can you explain the results of any product analysis or reverse engineering?

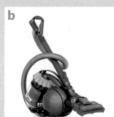

a b c d

Identifying the component parts of a product

When analysing existing products, it is good practice for an Engineer to be able to quickly categorise the various components that are used in products. By categorising the various components, you can quickly identify different systems that go towards creating the product and then seeing how those systems interrelate and how they work together to make the product function.

The different categories of components that you need to look at identifying are:
* component parts
* electrical components
* mechanical components
* materials and properties.

Task 4.2

Following are four pairs of images of a simple hairdryer. On the left are internal views of the product and on the right are views with the case/cover fitted.

Here is the list of component parts:
* case
* element cover
* dust cover
* heating element
* insulated copper wire
* screws
* wire flex
* speed switch
* heat switch
* fan
* electric motor.

Look at the following categories of components in the hairdryer. Sketch the hairdryers in your notebook and, using arrows, identify and label all the relevant parts (*some parts of the hairdryer will fit into more than one category*).

Component parts

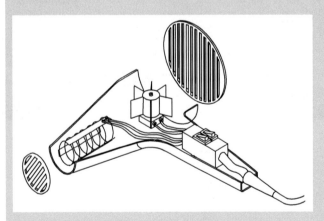 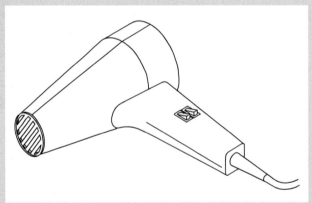

Electrical components

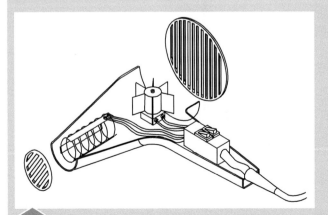 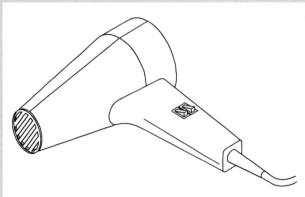

Task 4.2 *continued*

Mechanical components

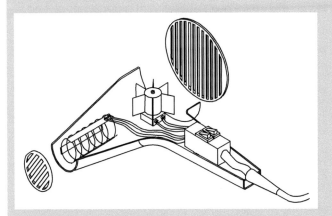

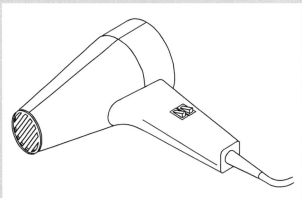

Materials and properties

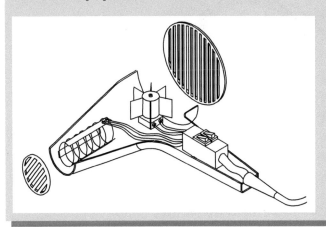

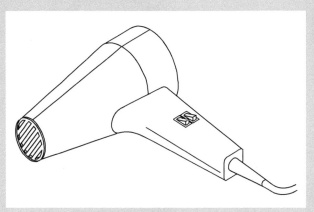

Task 4.3

Take a picture of a product from home or the internet and attempt to identify and label the:
* component parts
* electrical components
* mechanical components
* materials and properties.

Present your work on an A3 sheet. You can do this by hand or on a computer, or use both methods.

How products function

Engineers also need to understand how products function. Existing products tend to be made from lots of smaller component parts that all work together to create a functioning product. Sometimes the product will have moving parts/mechanisms and sometimes the product could be stationary but still have several different components … even just a finish (see Chapter 3 for more on finishes).

You have already learnt how to identify the different component parts of a product and can now label individual components in isolation.

Now you need to see how those components work **together** to create a working/functioning product and 'explain' how the component parts interrelate to create a functioning product.

Look at the diagram of the bicycle below. Can you see the component parts?

All the component parts of the bicycle below have been labelled with explanations of their function.

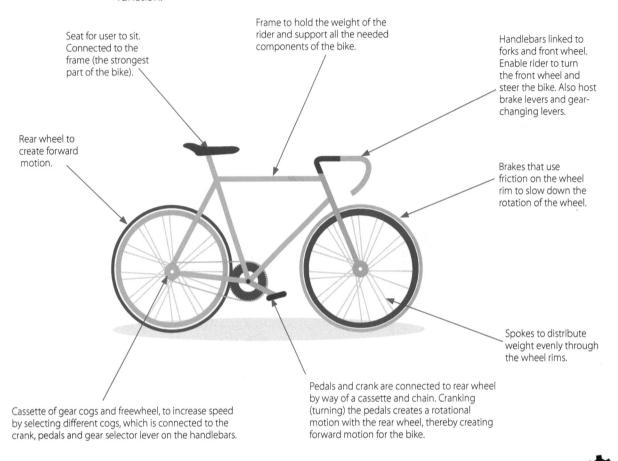

Seat for user to sit. Connected to the frame (the strongest part of the bike).

Frame to hold the weight of the rider and support all the needed components of the bike.

Handlebars linked to forks and front wheel. Enable rider to turn the front wheel and steer the bike. Also host brake levers and gear-changing levers.

Rear wheel to create forward motion.

Brakes that use friction on the wheel rim to slow down the rotation of the wheel.

Spokes to distribute weight evenly through the wheel rims.

Cassette of gear cogs and freewheel, to increase speed by selecting different cogs, which is connected to the crank, pedals and gear selector lever on the handlebars.

Pedals and crank are connected to rear wheel by way of a cassette and chain. Cranking (turning) the pedals creates a rotational motion with the rear wheel, thereby creating forward motion for the bike.

Task 4.4

1. From the list below, identify the component parts of the drill shown in the image on the right.
2. Copy the image of the drill into your notebook and using arrows (where needed) draw links between the component parts that directly work together.
3. Write down and describe how the component parts interrelate to create a functioning drill.

Component parts:
* battery pack
* variable resistor
* trigger
* electric motor
* gearing
* chuck
* drill bit
* copper wires
* wire connectors.

Key term

Interrelate: work together.

Task 4.5

On a separate A3 sheet of paper, select a product (drill, hairdryer, bicycle or other) to analyse.

Place a picture of your chosen product in the middle of your A3 page and draw links between the main component parts.

Describe how the component parts interrelate to perform a function.

Try to choose a product/picture where you can see the component parts.

Try to choose a simple product.

Use lots of annotation to explain/describe the functions.

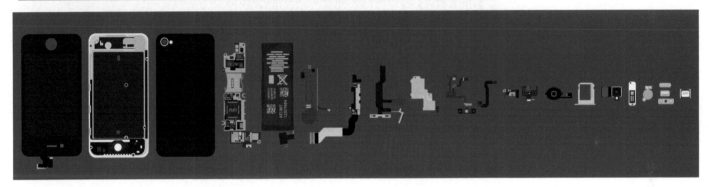

↑ *The disassembled components of a mobile phone.*

5

Analysing and Designing Products Against a Brief

In this chapter you are going to:
- → Analyse a design brief
- → Identify key features of a design brief
- → Identify key features of existing products that relate to a design brief
- → Begin to develop solutions to a design brief using your learnt communication skills.

This chapter will cover the following areas of the WJEC specification:

Unit 1 LO3 Be able to propose design solutions	
AC3.1 Develop creative ideas for engineered products	Creative ideas: identify features of other engineered products; generate ideas; explore implementation of ideas
Unit 3 LO4 Be able to solve engineering problems	
AC4.3 Analyse situations for engineering problems	Analyse: filter information; synthesise information; identify salient points; identify requirements
AC4.4 Propose solutions in response to engineering problems	Propose solutions: communication; logical structure

Introduction

We have already learnt how Engineers need to communicate effectively with both non-Engineers and other Engineers. Sometimes an Engineer would also need to communicate with customers, clients and colleagues within their company. These groups of people would generally be the ones who give engineers the tasks they need to perform. The formal name of these 'given tasks' is a design brief.

Design briefs

A design brief can come in many forms. Sometimes it could be a casual discussion with the Engineer on what would be needed, and then left to the Engineer to propose solutions. At other times an Engineer might find themselves with a formally written document, outlining lots of specific areas that need to be addressed. Either way, it is still the Engineer's job to be able to analyse a brief effectively, identify the key, relevant features of the brief and then propose a number of solutions that would satisfy/meet the brief.

Following is an example of a written design brief. This could be a typical series of statements that an Engineer would receive from clients/colleagues.

Design brief

A large manufacturer of bicycles and bicycle add-ons is looking to produce a new type of cycle specifically to be used for shoppers who are high-street shopping. The solution could be a bicycle, tricycle or any cycle combination that would be best for the shopping task. The cycle solution would need to be made from existing materials that the manufacturer currently uses, to ensure that production of the new product is feasible. There must be sufficient space to transport shopping items (clothing, groceries, etc.) as well as having an option for security. The new cycle solution would need to be used by a range of people with different heights and sizes. The target market would also need to be able to afford to purchase the new cycle, so the cost of manufacture would need to be considered.

In this brief there are key features that an Engineer would need to produce successful solutions. However, there are also pieces of information that may not be that useful when developing solutions. So, let's highlight all the key features that can be seen.

Key terms

Feasible: manageable/possible.
Key features: relevant pieces of information.

Design brief with highlighted key features

A large manufacturer of bicycles and bicycle add-ons is looking to produce a new type of cycle specifically to be used for shoppers who are high-street shopping. The solution could be a bicycle, tricycle or any cycle combination that would be best for the shopping task. The cycle solution would need to be made from existing materials that the manufacturer currently uses, to ensure that production of the new product is feasible. There must be sufficient space to transport shopping items (clothing, groceries, etc.) as well as having an option for security. The new cycle solution would need to be used by a range of people with different heights and sizes. The target market would also need to be able to afford to purchase the new cycle, so the cost of manufacture would need to be considered.

Once the identifiable key features have been highlighted, you can rewrite the design brief in a condensed format, or even write a bullet-point list of things that the client/colleague has asked you to produce.

Key term

Condensed: reduced so anything not required is taken out.

Condensed brief

Develop a new cycle for shoppers. It can have any cycle combination, made from current-use materials and is easy to manufacture. It needs storage space, needs to be adjustable, lockable and has to be low cost.

And now let's try a very simple bullet-pointed list of key features.

Bullet-point brief list

- Cycle for shoppers
- Two, three or four wheels
- Existing materials (manufacturer)
- Easy to make (manufacturer)
- Storage
- Adjustable
- Possibly lockable
- Low cost.

Once the list has been compiled, you can then look at prioritising it, so the most important task/areas to look at can be dealt with first. This also helps to identify what resources might be needed and when to obtain those resources.

Prioritised bullet-point brief list

Order of importance:
1 = most important
8 = least important

1. Cycle for shoppers
2. Low cost
3. Storage
4. Adjustable
5. Existing materials (manufacturer)
6. Easy to make (manufacturer)
7. Two, three or four wheels
8. Possibly lockable.

As you can see, having the knowledge and skill to extract the key features of a design brief can benefit an Engineer when trying to identify exactly what is needed in an engineered solution.

Top tip

Having an easily identifiable list of key features can save time and reduce confusion. It also allows the Engineer to prioritise tasks and resources.

Task 5.1

Read the following brief. Using your new knowledge of how to analyse design briefs, pick-out the key features of the brief and then, in your notebook, write a shorter, more concise version. Once you have completed that exercise, write a prioritised list of the key points with the 'most important' first and the 'least important' last.

Once complete, compare the prioritised list to the original brief.

Which would be easier to work with and why?

Design brief

A large car manufacturer is looking to develop a new 'universal' steering wheel system for its new range of cars. The new 'customisable' car industry has proven to be a success and the manufacturer wants to be part of the growing trend.

It is exploring the possibility of each customer having their own bespoke, individually designed steering wheel that they can quickly un-clip and re-use in their other car models if needed.

The new steering wheels must be fully customisable, with the option to swap-out designs, styles and accessories, with a view towards developing more 'options' in the future. The new steering wheels must have a universal fitting system that can be used on all other car models from the same manufacturer. The new wheels must be easy to put on and take off the steering column with no more than one button or lever to perform this function.

You must also look at how much it would cost as well as the potential for improved security.

Priorities
1.
2.
3.

Creating solutions from a design brief

Once you have developed the ability to identify and extract the key features of a brief, you can now use them to start creating solutions that go towards satisfying it.

Task 5.2

This task consists of two different design briefs.
- Identify the key features of each brief.
- Draw a solution using a recognised drawing technique.
- Annotate and describe the key features (including material choice and why).

Design brief 1

A large furniture manufacturer has asked you to design a seating solution for pupils in schools. The solution should fit all pupils in a secondary school. It must be comfortable to sit on and be extremely durable. The school's contract for this new seating solution does not come with a great deal of money, so your finances are limited. The seating solution will be placed in all new schools. All new schools have to conform to a green/environmental policy.

Design brief 2

Manchester United has asked you to come up with a solution to their coach seating problem for its training grounds. It wants somewhere for the coaches to sit when watching the team train. The solution will need to be movable. United wants the coaches to stay dry and be able to have an unrestricted view of the players. The solution will be placed on the training grounds and money is no object.

Top tip

When creating designs, use your drawing skills and stick to drawing in standardised formats, isometric, for example.

Top tip

The annotation of designs should also talk about the brief and how the key features are represented.

6 Design Specifications

In this chapter you are going to:
→ Discover what a design specification is
→ Find out what information needs to be included in a design specification
→ Learn how to construct a design specification
→ Find out how to use prior learnt knowledge to help write a design specification.

This chapter will cover the following areas of the WJEC specification:

Unit 1 LO3 Be able to propose design solutions	
AC3.3 Produce design specifications	Design specifications: clear communication; demands/wishes; using prepared templates; using set criteria

Introduction

As you are aware, Engineers are often asked to create new solutions to existing problems. When most of the research work has been completed (analysing briefs, product analysis, reverse engineering, etc.), Engineers create a written document listing all the things the new solution will or may need in order to be successful. This list of 'success criteria' is known as a design specification.

Design specifications should only be written once relevant information has been gathered from the research and can be used to define the criteria for success. Once the design specification has been written, it can then be used as a guide for the Engineer to check progress against, as well as measure the developments and solutions against. The design specification is also a great tool to use when evaluating the success of your final outcome.

What do you need to include in a design specification?

Much like the product analysis process, a design specification needs to contain lots of different criteria that are relevant to the brief and the solution you are developing. These criteria can be written in headings and titles, such as what materials would be used, what it would look like, sizes and other appropriate criteria.

How should the criteria be written down?

Once you have completed the research you will understand what the solution must be or must have. You can then begin to write statements or points, explaining what the solution could be. A typical design specification point could be written like this:

> The solution must be made from a non-ferrous metal to ensure it will not corrode.

This specification point would likely be under a heading of **materials**.

What else makes a good design specification?

A hierarchy

A design specification should also contain a hierarchy. A hierarchy is essentially a list of things from the most important to the least important. By using a hierarchy in a design specification, an Engineer can identify which points should be concentrated on first and the available resources can be targeted at these things first.

Ways of splitting your specification up into a hierarchy

You could split all the specification points into two lists:
- an **essential** list of specification points, and
- a **desirable** list of specification points
with the essential list being the most important.

You could also assign criteria or numbers against each specification point, with a key explaining what each of the numbers or symbols mean, as shown on the right.

Qualitative and quantitative data

A good design specification should also contain qualitative and quantitative data. Qualitative and quantitative data are sets of data (information) that complement each other when either gathering or using information. In a design specification, both sets of data can be used as guides for success criteria and clear areas that can be used to measure your solution against.

So what is the difference between qualitative and quantitative data?

Qualitative data is:
- About things that may not be strictly measurable.
- About descriptions, feelings, opinions and emotional responses.

Quantitative data is:
- About things that can be measured and weighed.
- About facts and numbers.

Top tip

ACCESSFM could be a good starting point for design specification criteria.

Key
 1 = most important
10 = least important

OR

Key
★★★★★ = most important
      ★ = least important

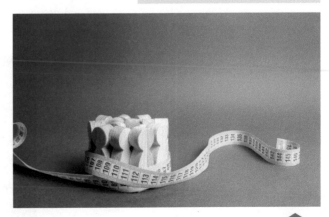

↑ *Quantitative data is about things that can be measured or weighed.*

Mobile phone

Qualitative data:
* How does it look?
* How does it feel?
* Do you have an emotional response?
* How do you feel when using it?

Quantitative data:
* What is it made from?
* How big is it?
* How heavy is it?
* How much does it cost?
* What is the scale of manufacture?

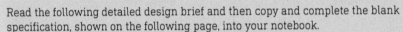

Top tip

Your measurable criteria don't necessarily have to be just numbers, such as sizes. Your measurable criteria could also include colours or even your end users' opinions.

Measurable criteria

A good specification should also contain points that are clearly measurable. This will allow you to successfully evaluate your solution against your written success criteria (specification).

Task 6.1

Read the following detailed design brief and then copy and complete the blank specification, shown on the following page, into your notebook.
* Remember to check the design brief for any key features.
* A specification template has been created for you to copy but you can create your own style of specification and include any of the criteria you have learnt about so far (hierarchy, qualitative/quantitative, etc.).
* One essential criterion has been written as an example.

Design brief

You have been asked by an engineering automotive company (e.g. Ford or Toyota) to develop a solution to the following problem:

With the advances in tyre technology, punctures are becoming less of a problem to the daily motorist. However, it is a problem that cannot be completely eradicated and motorists are still having to pull over on busy motorways and roads to change wheels. This is a time-consuming process and a certain amount of force is needed to lift the wheel in place (onto the bolts).

We would like you to develop a solution to make the lifting and placing of a spare wheel onto the raised car easier for motorists. It must be small enough to fit in the boot of a car and not take up any extra space. Low cost enough that it can be made cheaply and in the millions. Strong enough to support a spare wheel and also be simple to use.

Task 6.1 *continued*

SPECIFICATION

ENGINEERING DESIGN: ....

NAME: ....

ESSENTIAL CRITERIA:

Materials ....
It should be made from a metal with good tensile strength.
It should be made from a metal that is easy to fabricate.
The materials should be finished for anti-corrosion purposes.

DESIRABLE CRITERIA:

7

Evaluating Design Ideas

In this chapter you are going to:
- → Learn why we evaluate
- → Discover what an effective evaluation is
- → Find out how evaluations help us
- → Discover the different methods of evaluating
- → Learn how we can use different 'techniques' of evaluation.

This chapter will cover the following areas of the WJEC specification:

Unit 1 LO3 Be able to propose design solutions	
AC3.2 Evaluate options for design solutions	Evaluate: constraints; design requirements; fit for purpose; best fit; operating performance; reliability Evaluation techniques: total design model; SWOT analysis; advantages and disadvantages

Introduction

If a human being wants to progress and learn, they have to develop and use evaluation skills.

The first-time you crossed a road on your own you may not have realised at the time that you completed an internal evaluation of your performance. You would have asked yourself: How did I do? Did I judge the speed of the traffic correctly? Did I judge the distances correctly? Am I still alive?

By evaluating your performance, you would learn something that would allow you to perform better next time.

When working on projects it is good practice for Engineers to evaluate their performance, not only for the final solution, but also as the project progresses (correct materials, procedures, finishes, timings, etc.).

When you evaluate performance, you are able to look at your findings and use them to improve your performance and outcome when you are given your next project or brief.

You can evaluate on an *informal* basis by asking yourself questions, asking others their opinions or even visually checking outcomes against your success criteria. However, in this chapter we will be looking at methods of *formally* evaluating using existing methods and models.

Evaluating ideas

It can be difficult to start evaluating a series of ideas and solutions and even progress. Where do you start? What do you look for? What should you say?

One of the most common and successful ways of evaluating solutions is to evaluate against a design brief and specification (the specification being your very own list of success criteria).

The following are different but accepted methods of evaluating progress and outcomes of projects that an Engineer may be required to work on.

> **Top tip**
>
> You can also use set criteria (headings) to help you break down the task into easy-to-manage sections.

Best fit

Look at the ideas/solutions you have created. Compare them to each specification point. Which ideas match the requirements of the specification points best? Which idea meets the most specification points (could this be the best idea?). Which idea meets the least amount of points (could you stop developing this idea?). You could even do this in a table format, as shown below, to keep it formal.

	Idea 1	Idea 2	Idea 3
Specification point 1	✗	✓	✓
Specification point 2	✗	✓	✓
Specification point 3	✓	✗	✓
Specification point 4	✗	✓	✓
Total	**1**	**3**	**4**

You can see that using this method clearly shows the idea that best fits the specification (Idea 3) and would therefore be more likely to be the optimum choice to develop.

Design requirements/fit for purpose

By extracting the key features of a design brief, you are able to quickly identify the **design requirements** needed in your solution.

When you have identified the design requirements you can then compare your solution to the key features of the design brief and check to see if your solution is **fit for purpose**.

You can also perform this evaluation task by drawing a simple table and checking off the design requirements, as shown below.

Design requirements (design brief key features)	Solution outcome	Fit for purpose?
Must be portable	Yes, as it is lightweight.	✓
Must fit into one hand	Yes, as the size is small enough.	✓
Must have a range of colours	Yes, as it is made from ABS plastic.	✓
Must be easy to charge	No, as it needs a computer to charge.	✗

Evaluating with this method helps to identify areas that need addressing and you can go back to your solution and further develop it to meet all the design requirements.

Constraints

Look at your range of solutions and ideas. Which one is going to offer the most constraints? What limitations are you working with?

It might be that your workshop or manufacturing plant only has access to limited manufacturing equipment. Maybe you have a short time to create the solution or even need to learn new skills to be able to complete the project. A big constraint in industry is the overall cost of a project and whether or not some potential ideas are viable to run.

Think about all the constraints and ask yourself these questions:
- Will it be difficult to make?
- Is it too expensive to make?
- Are there enough facilities available to be able to make it?
- Do I have the skills to make it?
- Is there similar competition or is it too close a copy?
- Will the idea have to be compromised/developed too much to make it work?

Top tip

By answering these types of questions you will also have a good idea of which ideas would not be viable to continue with and would help narrow the list of potential solutions.

Reliability/operating performance

You can also evaluate the potential reliability and operating performance of your list of solutions. You can try to predict how well they will function during their working life, with the end users/target market and also in the environment they will be used. You could ask:
- Which one is likely to be the most reliable?
- Which will be the most functional?
- Will it continue to perform its function for a long period of time?
- Which one would the end user find easiest to use?

These questions can be answered by analysing the component parts of the solution such as materials (will it corrode?), sizes (too big or too small?), shape (comfortable to use/fit in its environment?), finish (easy grip, safe to use and so on?).

These questions will help you decide which solution will be the optimal solution for the project you are working on.

Key terms

Optimal: the absolute best.
Obsolescence: a product is created to become obsolete – old-fashioned or out of date.

However, although Engineers and Designers create new product solutions that are the optimal design, some products are intentionally designed to fail. Imagine a razor-blade that never went blunt or a vacuum cleaner that never lost its suction or broke-down. What would happen to the companies that made those products? Would they go out of business when the consumers stopped buying their products because they didn't need to? Operating performance and reliability of a solution also have to be measured by the needs of the company and client, which is why some products are designed to fail for different reasons such as:
- shelf-life (fashion/trends/technology)
- time (reliable for a set amount of time)
- serviceability (parts break and the product needs to be serviced).

In design, this is called obsolescence (create it to become obsolete).

Top tip

When evaluating reliability and operating performance of solutions, ALL the needs of the interested parties (customer, company, etc.) need to be taken into consideration.

Evaluation techniques

You can also evaluate using existing 'evaluation techniques'. These techniques are recognised evaluation models and are also a useful tool in the engineering/design industry to evaluate projects, outcomes and processes.

Total design model

The total design model encourages you to evaluate the WHOLE of the impact of the design project. It focuses on not only the end solution but also looks at evaluating the entire design process. This could mean that you evaluate how well you analysed the products or design brief, how well you developed your ideas, or even how well you used the tools and equipment. The total design model helps you to see the WHOLE process and allows you to write a very detailed evaluation of the entire project. You could then use lots of criteria and headings with which to evaluate.

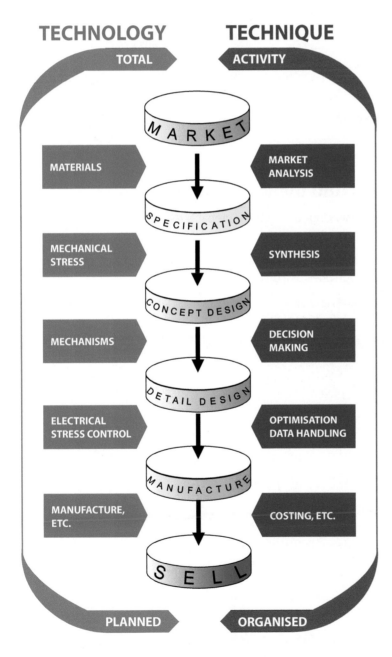

⬆ *A visual model of the 'total design model' with a visual guide of different areas of the design process. You can use this model to create criteria/headings for your evaluation process. (Adapted from Pugh, S. (1991) The Total Design.)*

SWOT analysis

A SWOT analysis is a useful evaluation tool as it can help you to discover opportunities that can be exploited when working on a project, the strengths of your project and will also identify any weaknesses your ideas may have. SWOT can also discover any threats your project may have by looking at competitor products.

Key term

SWOT: stands for:

- **Strengths:** identifies good points about your project/idea.
- **Weaknesses:** identifies areas of your project/idea that could make it fail.
- **Opportunities:** identifies areas of your project/idea that you could exploit.
- **Threats:** identifies areas that could cause trouble for the project/idea.

SWOT

STRENGTHS

List the:

- Key features that match the design brief
- Key features that match the specification
- Things the target market would like

WEAKNESSES

List the:

- Limitations of the idea
- Things the target market would not like
- Points on the specification it did not meet

OPPORTUNITIES

List the:

- Ways in which the idea could be improved/ developed

THREATS

List the:

- Other products in the market that are similar
- Extra resources needed to make it
- Extra money required to make it

Advantages and disadvantages

The next model of evaluating you can use is probably the most common and one you should be familiar with.

For your ideas you could list the advantages and disadvantages of each one using the specification and design briefs as a guide (success criteria).

By listing the advantages and disadvantages, in a similar table to the one below, you are able to see which ideas are best suited to carry forward and develop further.

Success criteria	Advantages	Disadvantages
Design brief		

Success criteria	Advantages	Disadvantages
Specification		

Unit 1 Submission

8

Introduction

Once you have reached this stage of the book you should now have all the knowledge needed to successfully complete and submit Unit 1.

The first seven chapters will have fully prepared you to create a portfolio of work that achieves all the Learning Objectives and success criteria in Unit 1. This includes having the ability and knowledge to access the top performance bands.

So what do you need to include in Unit 1?

In the **course structure** section of this book you will find a list of suggestions on content for Unit 1. Have a look at this section again to see what you will need to produce.

Alternatively, look at the table on page 84 to see what needs to be demonstrated, how you could demonstrate it and where to access the information needed to demonstrate your knowledge:

- The **first column** shows a list of Assessment Criteria from the specification that you will have to demonstrate for Unit 1.
- The **second column** offers suggestions on HOW you can demonstrate your knowledge (each school or college will have their own interpretations of the specification and may decide to demonstrate your knowledge in different, valid ways).
- The **third column** shows you in which chapters this book has covered the relevant areas.
- The **fourth column** can be used as a checklist to see if you are happy with your knowledge or need to re-visit the chapters to further increase your knowledge.

> **Link**
>
> For information on the course structure see pages 5–6.

What should I submit?

You can submit Unit 1 in any format that is easy for your school or college to work with, depending on the resources you have.

It can be submitted:
- on paper as a six–seven page portfolio, A3 or A4 (which would be best to show drawings and orthographic projections?)
- digitally
- any other format accepted by the WJEC.

Also ensure that the front page of your submission clearly displays the:
- unit number (1)
- centre number and name
- candidate number.

Lastly, don't forget to keep checking the **performance bands**. Look at the work you produced and ask yourself if you are achieving the performance band you think your work deserves. Don't forget, you can always add to your work providing you do not go above the seven-hour allotted time.

The following table shows you where in this book to find the relevant text for the skills required to demonstrate knowledge of the Assessment Criteria.

Assessment Criteria	Possible ways of demonstrating	Covered in chapters:	Happy with knowledge	Re-visit chapters
AC1.1 Identify features that contribute to the primary function of engineered products' features	• Identifying key features of a brief • Product analysis • Reverse engineering	3 Materials and Properties 4 Identifying Features of Working Products 5 Analysing and Designing Products Against a Brief		
AC1.2 Identify features of engineered products that meet requirements of a brief	• Identifying key features of a brief • Product analysis • Reverse engineering	3 Materials and Properties 4 Identifying Features of Working Products 5 Analysing and Designing Products Against a Brief		
AC1.3 Describe how engineering products function	• Product analysis • Reverse engineering • Generation of ideas (isometric) • Development of ideas (isometric)	3 Materials and Properties 4 Identifying Features of Working Products		
AC2.1 Draw engineering design solutions	• Generation of ideas (isometric) • Development of ideas (isometric) • Final design (isometric) • Final design (CAD) • Orthographic projection (3rd angle)	1 Engineering Drawings 2 Communicating Design Ideas 3 Materials and Properties 4 Identifying Features of Working Products		
AC2.2 Communicate design ideas	• Generation of ideas (annotations/explanations) • Development of ideas (annotations/explanations) • Final design • Orthographic projection (3rd angle)	1 Engineering Drawings 2 Communicating Design Ideas 3 Materials and Properties 4 Identifying Features of Working Products 7 Evaluating Design Ideas		
AC3.1 Develop creative ideas for engineered products	• Generation of ideas (annotations/explanations) • Development of ideas (annotations/explanations/links to existing products) • Final design	1 Engineering Drawings 2 Communicating Design Ideas 3 Materials and Properties 4 Identifying Features of Working Products 5 Analysing and Designing Products Against a Brief 7 Evaluating Design Ideas		
AC3.2 Evaluate options for design solutions	• Generation of ideas (annotations/explanations/justification of final choice) • Development of ideas (annotations/explanations/justification of changes) • Evaluation of ideas/final idea/development changes	2 Communicating Design Ideas 4 Identifying Features of Working Products 5 Analysing and Designing Products Against a Brief 7 Evaluating Design Ideas		
AC3.3 Produce design specifications	• Design specification	6 Design Specifications		

Managing and Evaluating Production

9

In this chapter you are going to:
- → Find out how to accurately interpret technical information
- → Discover how to identify resources needed to produce the solution
- → Learn how to organise information so it can be used effectively
- → Learn how to sequence activities correctly.

This chapter will cover the following areas of the WJEC specification:

Unit 2 LO1 Be able to interpret engineering information	
AC1.1 Interpret engineering drawings	Interpret: symbols; conventions; information; calculation Sources: sketches; drawings; design specifications
AC1.2 Interpret engineering information	Engineering information: data charts; data sheets; job sheets; specifications; tolerances
Unit 2 LO2 Be able to plan engineering production	
AC2.1 Identify resources required	Resources: materials; equipment; tools; time
AC2.2 Sequence required activities	Sequence: prioritise activities – which are needed before something else can be done; within designated parameters; consideration of resources available; contingencies

Introduction

When Engineers are asked to produce a new product or make a prototype of the new solution they have developed, they start with technical information from a page or digital file. Engineers have to accurately interpret this information and organise data in a way that is easy to use, follow and apply. By spending time organising this information before starting the production processes, Engineers are less likely to make mistakes and waste resources, and are more likely to produce a working prototype to a high standard.

In this chapter you are going to look at how to interpret and use the information you are given as Engineers, and use it to plan and evaluate the production of a product/solution. You will also look at how you could evaluate the outcome of the solutions.

Interpreting engineering drawings (orthographic projections)

When starting the process of producing a prototype (or any other production process), Engineers will most likely have an orthographic projection (engineering drawing) with which to work. It might be that the Engineer themself has produced the engineering drawing or they might have been given it as part of a 'working group'.

If the engineering drawing has been completed to a high standard it will have lots of information on it that the Engineer can use to start planning the project.

↑ *Engineering drawings contain lots of information.*

Let's look at an engineering drawing for several parts for a modern desk lamp. As you are now able to produce engineering drawings yourself, use your learnt knowledge to help you in this process.

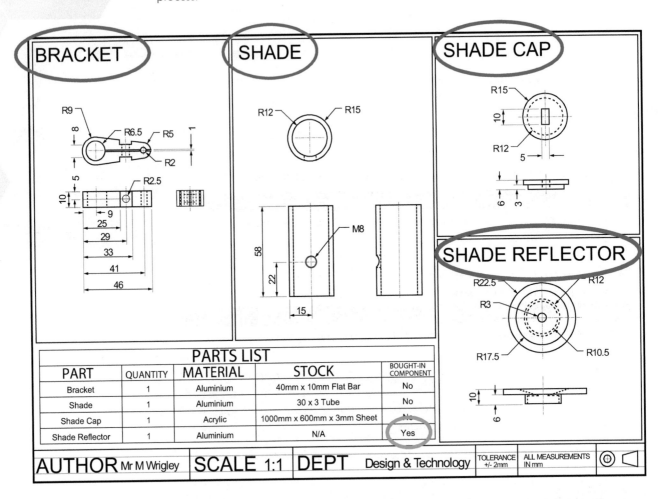

PARTS LIST				
PART	QUANTITY	MATERIAL	STOCK	BOUGHT-IN COMPONENT
Bracket	1	Aluminium	40mm x 10mm Flat Bar	No
Shade	1	Aluminium	30 x 3 Tube	No
Shade Cap	1	Acrylic	1000mm x 600mm x 3mm Sheet	No
Shade Reflector	1	Aluminium	N/A	Yes

AUTHOR Mr M Wrigley · SCALE 1:1 · DEPT Design & Technology · TOLERANCE +/- 2mm · ALL MEASUREMENTS IN mm

↑ *This drawing shows the four listed parts (circled in red) and how the shade reflector is a bought-in component (circled in green).*

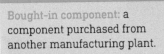

Key term

Bought-in component: a component purchased from another manufacturing plant.

Firstly, you can see that on this drawing there are four different parts. However, if you look at the 'Parts List' you can see that the 'shade reflector' is actually a bought-in component and would not need to be made by yourself. So, already you are starting to identify the relevant information. Now you only need to concentrate on the relevant parts and all the work that is produced to coordinate the project will be worthwhile.

In the following drawing you can next identify the overall dimensions and sizes of the parts you need to manufacture. By looking at the dimensions and materials they are made from you can identify **how much** material would need to be purchased and what stock-form they can be purchased in. The 'Parts List' also tells you how many of each part is needed (quantity), which would also have an impact on what materials you may need to purchase.

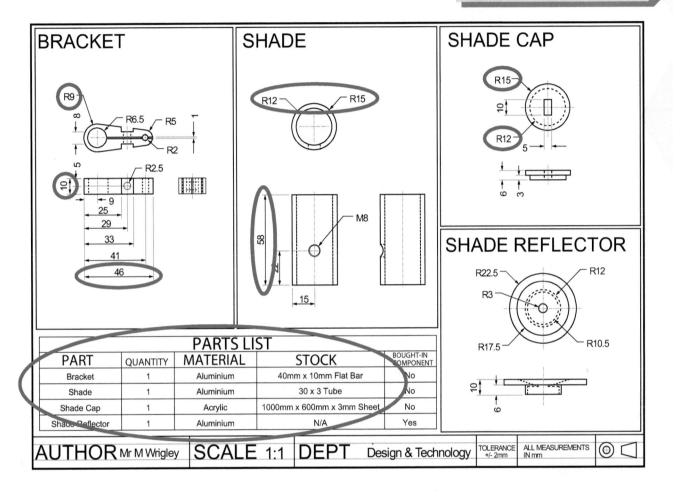

PARTS LIST				
PART	QUANTITY	MATERIAL	STOCK	BOUGHT-IN COMPONENT
Bracket	1	Aluminium	40mm x 10mm Flat Bar	No
Shade	1	Aluminium	30 x 3 Tube	No
Shade Cap	1	Acrylic	1000mm x 600mm x 3mm Sheet	No
Shade Reflector	1	Aluminium	N/A	Yes

| AUTHOR Mr M Wrigley | SCALE 1:1 | DEPT Design & Technology | TOLERANCE +/- 2mm | ALL MEASUREMENTS IN mm | |

↑ *This drawing shows dimensions, materials and how much stock-form you need to buy to complete the project.*

The following drawing shows you the amount of tolerance each part can have. This may have an impact on what workshop processes you decide to choose to complete your parts, as different processes have different levels of accuracy. For example, would you use a metal file to get a flat surface or a milling machine? What about a hacksaw or metal bandsaw to cut materials to size? Each decision you make at this stage will have an impact on the quality of your outcome, so gathering as much information as you can and making 'informed' decisions will greatly benefit the project as a whole.

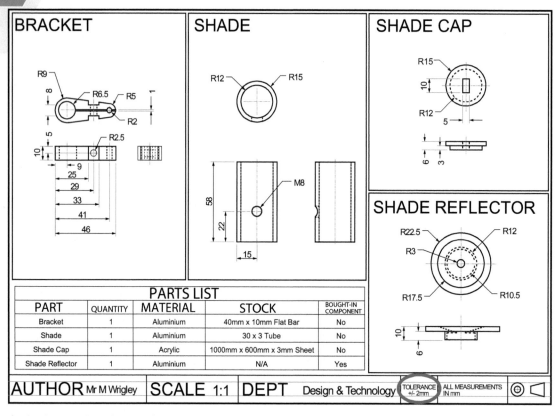

PARTS LIST				
PART	QUANTITY	MATERIAL	STOCK	BOUGHT-IN COMPONENT
Bracket	1	Aluminium	40mm x 10mm Flat Bar	No
Shade	1	Aluminium	30 x 3 Tube	No
Shade Cap	1	Acrylic	1000mm x 600mm x 3mm Sheet	No
Shade Reflector	1	Aluminium	N/A	Yes

| AUTHOR Mr M Wrigley | SCALE 1:1 | DEPT Design & Technology | TOLERANCE +/- 2mm | ALL MEASUREMENTS IN mm |

⬆ *This drawing shows how much tolerance you can have for each part.*

Creating cutting lists, job sheets and sequencing

Now the main parts of an engineering drawing that will have the most impact on producing a prototype have been identified, you need to start looking at ways of organising this information so it can be used easily.

Cutting lists

A cutting list is a simple table showing how each part is to be cut and at what size (from the correct material/stock-form).

Using the information gained from interpreting the engineering drawings, you can now produce a cutting list. Look at the cutting list below and notice how the information is organised to be easily read and used, as well as including the tools and process needed to cut the material to size.

CUTTING LIST				
Part	Material	Stock-form	Cut to size	Tools/equipment needed
Bracket	Aluminium	50mm × 10mm flat bar	20mm	Hacksaw/metal bandsaw
Shade	Aluminium	30mm × 24mm round tube	60mm	Hacksaw/metal bandsaw
Shade cap (top)	Acrylic	3mm sheet	R30mm	Laser-cutter/bandsaw
Shade cap (bottom)	Acrylic	3mm sheet	R30mm	Laser-cutter/bandsaw

⬆ *The information in this table allows you to start sourcing the correct materials and begin the process of cutting them to size.*

Job sheets and sequencing

A job sheet is similar to a cutting list as it can come in a 'chart' format and has information that can be used to help create prototypes efficiently and accurately. Job sheets, however, are designed to show what 'jobs' need to be completed for the project and also in what order they need to be completed.

To create a part (from the engineering drawing) you would need to access the correct material, cut it to the correct size and then go through a series of processes that use tools and equipment to make sure the part is shaped and completed to the correct dimensions (that fall within tolerance). A job sheet can list all the tools and operations needed to complete the part successfully as well as including other relevant areas such as risk, time and quality control.

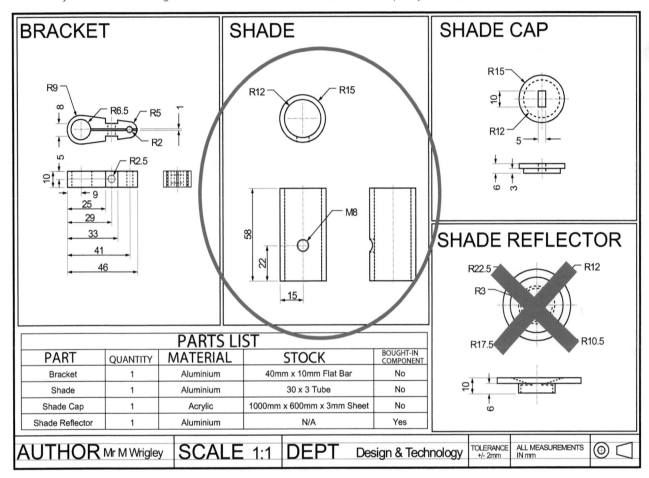

↑ *This drawing shows how you can focus on one part ('SHADE' in this case) and identify what processes would need to be completed.*

By looking at the part 'SHADE' in the engineering drawing above you can identify the following processes:

- To create the shade, the aluminium tube would need to be:
  1. Marked out.
  2. Cut.
  3. Centre punched.
  4. Drilled.
  5. Have a thread tapped.

- To perform these tasks you will need:
  1. steel rule/Vernier calliper, scriber, engineer's blue, V-blocks
  2. metal vice, hacksaw/metal bandsaw
  3. hammer (ball-peen), centre punch, V-blocks
  4. pillar drill, 7mm HSS twist-drill bit
  5. M8 tap, tap wrench, metal vice.

Once you have interpreted this information from the engineering drawing you can put it into a format that will be very useful when creating the prototype.

> **Top tip**
>
> To identify what processes are needed to complete a job sheet you can go back to your engineering drawing and look at the orthographic projections of the part. This will help you to identify which processes need to be completed (see page 89).

However, it is **very important** to note that the jobs and processes you list should be written in a workable **sequence**. This means that all processes should be completed in order. For example, when you look at the processes that have been identified for the 'SHADE' you can see that it needs to be drilled. Before the drilling takes place, however, you would need to make sure you are drilling in the right place. Therefore, marking-out, cutting and centre punching are three processes that would need to be completed BEFORE drilling takes place. Getting these tasks in the correct order is known as sequencing. Sequencing would also need to be correctly applied to the job sheet.

How you lay-out your job sheet is up to you. Maybe the company you work for has a standardised format that you work to. The main function of any good job sheet is to make sure you have quick and efficient access to the relevant information needed to complete the task successfully.

Following is a job sheet that has been created for the 'SHADE' part in the engineering drawing example (see page 89). Look at how the chart has been created and what titles/columns have been added to give more information.

(see page 89)

JOB SHEET
Part: SHADE (aluminium round tube 30mm × 24mm)

Part	Material and stock-form	Process and sequence	Tools/equipment needed	Risk level	Time	Health and safety considerations	Quality control
SHADE	Aluminium round tube 30mm × 24mm	Step 1: Mark-out	Steel rule/Vernier calliper, scriber, engineer's blue, V-blocks	Low	10 mins	n/a	Check accuracy of marks
		Step 2: Cut	Metal vice, hacksaw	Medium	5 mins	Sharp hacksaw blade	
		Step 3: Centre punch	Hammer (ball-peen), centre punch, V-blocks	Low	5 mins	Strike centre punch squarely Secure round tube in V-block	
		Step 4: Drill	Pillar drill, 7mm HSS twist-drill bit	Medium–high	5 mins	Correct set-up of pillar drill Wear PPE	Check set-up and drill RPM
		Step 5: Tap a thread	M8 tap, tap wrench, metal vice	Low	10 mins	Sharp teeth on tap	Keep tap vertical to ensure straight thread

↑ An example of a job sheet that includes sequencing as well as risk assessing, quality control and time. The layout is easy to understand and use in a workshop environment.

Using data sheets

When creating a prototype in a workshop environment you will more than likely be expected to use machinery to successfully and accurately create a prototype. You should also be tutored on the correct procedure for using the various machines that would include the correct set-up of the machine, the correct way of using it, as well as what health and safety precautions you would need to incorporate. However, many machines come with interchangeable cutting tools (e.g. drill bits) for when using different materials on the machine. This is where you need to start using data sheets/charts that are industry standard guidelines or specifications set out by the machine manufacturer.

Using the correct settings and speeds for the machines will ensure:
• a good-quality finish
• longer working life of the part/machine (no breaking)
• safer operation (user not being hurt).

By not using the correct data you could:
• destroy your work (have to start again)
• break parts or all of the machine
• injure yourself or others.

Following is an example of a data chart for a centre lathe. It contains useful information such as what RPM is needed when using different sizes of materials and different materials. Other useful information on RPM when performing specialist operations such as parting and knurling is shown in the Top tip on the right.

GUIDELINE RPM FOR CENTRE LATHE					
Material diameter		Material			
Inches	Millimetres	Aluminium	Brass	Mild steel	Stainless steel
½	12.7	1,400	1,200	1,000	600
1	25.4	700	600	500	300
1½	38	500	400	300	200
2	50.8	350	300	250	150
2½	63.5	280	250	220	120
3	76	225	190	160	100

Top tip

Knurling and parting operations should only be run at 100RPM maximum.

Gantt charts

Another big resource for many Engineers, whether working for a company or even trying to complete a prototype in school or college, is TIME.

Time (along with cost) is one of the main constraints Engineers have when starting projects. Being able to organise time effectively is of great benefit when trying to manage projects. Having the ability to organise time effectively for projects means time is not wasted, money is saved and deadlines will be met.

Key term

Constraint: limitation.

Gantt charts (invented by Henry Gantt 1861–1919) are widely used by Engineers and in industry to organise time effectively. 'Time', when producing a job sheet, has already been discussed. However, having a chart that organises time, which can be scanned and understood quickly, is a very effective way of managing time.

Following is an example of a Gantt chart. This chart has been made for the engineering drawing example in this chapter. For this example you are concentrating on the 'SHADE' part only.

Part: SHADE											
Process	**Total time 1 hour (5-minute segments)**										
Marking-out	██	██									
Cutting			██								
Centre punch				██							
Drill					██						
Tap a thread						██	██				

← *The process only took 35 minutes to complete so the Gannt chart shows there is more time available for you to start other processes.*

By quickly looking at the Gantt chart on page 91, you can see that the processes for the 'SHADE' part will take a total of 35 minutes, with marking-out and tapping a thread taking the longest to perform. With this information in mind you can start identifying areas in the workshop or manufacturing facility that could have bottlenecks and slow production. This information will help you organise your time more effectively.

Task 9.1

Now you understand the importance of organising relevant information, and are able to use it quickly and efficiently, you can try producing your own job sheet.

Look at the following engineering drawing. For the part 'BRACKET', create your own job sheet.

You MUST include the following:
* titles
* materials or stock-form
* sequence of tasks (jobs)
* equipment needed
* time.

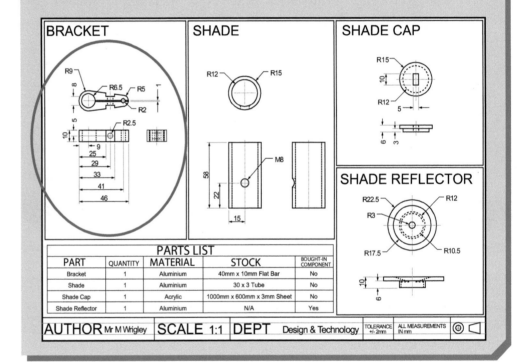

Health and Safety in the Workshop

In this chapter you are going to:
➔ Discover why health and safety in the workshop environment is needed
➔ Learn about risk assessments
➔ Understand health and safety signs (shapes and colours)
➔ Learn about wearable protection
➔ Learn how to use data.

This chapter will cover the following areas of the WJEC specification:

Unit 2 LO3 Be able to use engineering equipment	
AC3.1 Use tools in production of engineering products	Health and safety: awareness and application of health and safety practices
AC3.2 Use equipment in production of engineering products	Health and safety: awareness and application of health and safety practices
Unit 2 LO4 Be able to use engineering processes	
AC4.1 Use engineering processes in production of engineered products	Health and safety: awareness and application of health and safety practices

Introduction

In this chapter you are going to look at health and safety in the workshop environment. All Engineers have to understand and use a large range of tools and equipment during their career, whether it is starting out and going through training, or working in an office environment and specifying different pieces of equipment to be used for different processes. All Engineers also need to understand the health and safety procedures that have to be put in place to ensure safe working practices for the people involved.

When Engineers work with tools and machinery in a workshop environment, they need to understand the process of being as safe as possible.

When working on a machine that spins sharp/heavy pieces of metal at 2,000 revolutions per minute (RPM), welding steel at 10,000° Celsius or even producing toxic fumes from acid etching circuit boards, you need to make sure you can walk away from the process and equipment perfectly safe, with a quality outcome product.

This is why Engineers need to understand and apply health and safety procedures for each task that they perform.

Risk assessments

A risk assessment is the analysis of the risks involved when using equipment or performing a process.

Imagine making a cup of tea. Do you need to be aware of any dangers? What about the mixture of water and electricity? Would you be aware of the heat generated by the boiling water? What would you do to keep yourself safe and ensure you end up with a nice cup of tea? Identifying risks and putting procedures in place to keep yourself (and others) safe is called a risk assessment.

The five-step risk assessment

One of the most common (and recognised by industry) methods for preparing risk assessments is the five-step method:

Step 1. Identify the hazards.

Step 2. Who may be harmed and why?

Step 3. Evaluate risk and choose precautionary control measures.

Step 4. Record (write down) your findings.

Step 5. Review and update when needed.

Following is an example of a risk assessment, using these five steps, for using a bench/pillar drill.

RISK ASSESSMENT FOR BENCH/PILLAR DRILL (*SCHOOL WORKSHOP*)		
Identify the hazards	**Who may be harmed and why?**	**Evaluate risk and choose precautionary control measures**
1. Workpiece spinning on drill bit.	1. User cutting hand on rotating workpiece.	1. Use machine guard to reduce risk of rotating workpieces contacting user.
2. Workpiece flying off machine.	2. User being hit by flying workpiece.	2. Use machine guard to reduce risk of user being hit.
3. Swarf flying up.	3. Swarf flying into user's eyes.	3. Wear goggles to reduce risk of eye injury.
4. Long hair/loose clothing tangling in rotating parts.	4. User being dragged into rotating parts.	4. Wear apron and tie down loose hair/clothing to reduce risk of entanglement.

Note:

All manufacturer guidelines should be used when this machine is in operation.

Having completed the first three steps, the fourth would be to produce a document (e.g. the above table) that would be kept on record and readily available for the users of the machine (bench/pillar drill) to access and read. Step 5 would be done if there are any changes made to the machine (or law). Quite often, risk assessments are reviewed and updated on an annual (yearly) basis to make sure the procedures used are still up to date.

Key term

Swarf: small pieces of metal.

Signage

Signage is the word used for all the signs that you may see in a workshop environment. Knowing how to translate and understand the signs in a workshop is vital when dealing with potentially dangerous equipment and processes. By ignoring signage you could be in danger of harming yourself or others. Do you know the signs used to warn you that someone may be arc welding in your workshop? What would happen to your eyesight if you did not understand the signs and just popped your head into a welding booth to say hello? Could you lose the use of your sight temporarily or even permanently? This is why all Engineers should be aware of the signage that is displayed in any workshop environment.

Safety signs tend to come in different shapes and colours that have specific meanings attached to them. The following table will give you the description of each shape and colour and what they mean.

Sign	Meaning	Shape	Colour
	Mandatory sign: specific instruction on behaviour	Round	White border, blue background, white pictogram
	Warning sign: giving warning of hazards or danger	Triangular	Black border, yellow/ orange background, black pictogram
	Prohibition sign: prohibiting behaviour and/or actions	Round	Red border, white background, black pictogram
	No danger: information on emergency exits, first aid, emergency stop, etc.	Square or rectangular	White border, green background, white pictogram

Quite often, signs will come with some instruction in the form of text to reinforce the message:

Some areas of work, such as construction sites or workshops, may have multiple instructions that may have to be applied at the same time. Following is an example of a sign with multiple instructions for visitors when arriving at a construction site:

Rhybudd safle adeiladu

Warning construction site

Warning sign

Dim personau heb awdurod

No unauthorised persons

Prohibition sign

Rhaid gwisgo dillad gwelededd uchel

High visibility clothing must be worn

Mandatory sign

Rhaid gwisgo esgidiau diogelu

Protective footwear must be worn

Mandatory sign

Rhaid gwisgo helmed diogelwch

Safety helmets must be worn

Mandatory sign

Top tip

You can find a list of common signs that you **may** find in a workshop environment on the following two websites:
- free signage UK:
  http://www.freesignage.co.uk
- Online Sign:
  http://www.online-sign.com/

Mandatory signs

Personal protective equipment (PPE)

Rhaid gwisgo menig diogelwch
Safety gloves must be worn

Rhaid gwisgo gorchudd llygaid
Eye protection must be worn

Rhaid gwisgo offer diogelu clustiau
Ear protection must be worn

Rhaid gwisgo offer dillad diogelu
Protective clothing must be worn

Rhaid gwisgo mwgwd weldio
Wear a welding mask

Rhaid gwisgo mwgwd wyneb
Wear face guard

Other mandatory signs

Defnyddio gard
Use guard

Diffoddwch pan nas defnyddir
Turn off when not in use

Warning signs

Perygl
Caution

Rhybudd sioc drydanol

Caution electric shock risk

Perygl paladr laser
Caution laser beam

Rhybudd gwasgu dwylo

Warning crushing of hands

Perygl nwy cywasgedig

Danger compressed gas

Lefelau sŵn uchel
High noise levels

Prohibition signs

Peidiwch â
chyffwrdd
Do not touch

Peidiwch â chyffwrdd pan
fydd yn symud
Do not touch
when in use

Peidiwch ag iro na glanhau
pan fydd yn symud
Do not oil or clean
when in use

Peidiwch
â rhedeg
Do not run

Dim fflamau
noeth
No naked
flames

Dim bwyd
na diod
No food
or drink

No danger signs

Botwm
argyfwng
Emergency
stop

Golchi llygaid
mewn argyfwng
Emergency
eye wash

Safle cymorth
cyntaf
First aid
station

Allanfa
argyfwng
Emergency
exit

Using data sheets

When using tools and machinery you will be given instruction on their correct and safe use by your tutor. However, when using machinery such as pillar drills, milling machines and centre lathes you will have a number of variables such as cutting tool sizes or diameters and types of materials that would change depending on what job, project or operation is being performed. Because all these aspects tend to change frequently, all the machine settings would also need to change to match the current job.

For example, one day you could be milling a small aluminium bracket on the vertical miller with a 5mm diameter cutting tool, then the following day you could be trying to mill a large slab of stainless steel with a 12mm diameter cutting tool. In this situation you would have to adjust the speed of the cutter (RPM) as well as how fast (feed) the cutting tool would mill the work.

This is where good Engineers use data sheets/charts that are either developed by other machine users or come direct from the manufacturer. There are mathematical formulae that can be used to determine the speed and feed rates of all machines, depending on material properties and cutting tool sizes; however, there are many data guideline sheets that will give you a good idea of the speed and feed rates needed for your current job.

Following is an example of a data sheet chart for a pillar drill. See how it has 'Extra notes' at the bottom, giving advice and guidelines on specialist operations such as counterboring and countersinking.

GUIDELINE RPM FOR PILLAR DRILL					
Drill bit diameter		Material			
Inches	Millimetres	Aluminium	Brass	Mild steel	Stainless steel
¼	6.35	1,200	1,200	750	500
⅜	9.5	900	600	500	375
½	12.7	600	500	250	250
⅝	15.8	360	400	200	150
¾	19	300	300	150	120
1	25.4	180	200	100	80

Notes:
Counterbore bits and operations should run at 200RPM maximum.
Countersink bits and operations should be run at 300RPM maximum.

By utilising these data sheets every time you set up a machine for a job, you will reduce the risk of the machine breaking, the cutting tool breaking or the workpiece being ruined and, above all, you will ensure the safety of the operator … you.

COSHH

COSHH awareness and training is a needed set of skills that Engineers have to be aware of. In a workshop environment, Engineers will be working with some substances that could be hazardous to your health and would need to be worked with, handled and stored in a safe and secure place.

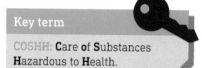

Key term

COSHH: **C**are of **S**ubstances **H**azardous to **H**ealth.

Substances that fall under the remit of COSHH include:
* chemicals
* products containing chemicals
* fumes
* dusts
* vapours
* mists
* nanotechnology
* gases and asphyxiating gases
* biological agents (germs). If the packaging has any of the hazard symbols on it then it is classed as a hazardous substance
* germs that cause diseases such as leptospirosis or legionnaires' disease and germs used in laboratories.

(Source: HSE (2019) 'What is a "Substance Hazardous to Health"?', http://www.hse.gov.uk/coshh/basics/substance.htm)

If any of the substances listed above are in a workshop you are working in or you are going to use any of the listed substances, then you must follow the government guidelines on their correct use and storage.

↑ *Dangerous fumes come under the remit of COSHH.*

As you can see, not all the COSHH substances listed would be found in a workshop's environment. However, here are some examples of substances that could be found in a workshop that would have to adhere to COSHH guidelines:

- paint
- varnish
- undercoat
- thinners
- solvents
- adhesives
- wood finishes (wax, etc.)
- acids
- fumes (welding, spray booths, etc.)
- dust particles (linisher/belt sanders/sanding)
- *any other substances that fall under COSHH guidelines.*

Other areas of safety guidelines in a workshop environment

Finally, there are other organisations that deal with health and safety in workshop environments. These organisations employ experts in their field and develop guidelines (not rules) on safe working practice from the setting-up of workshop spaces (distances between each machine, etc.) to how to safely use and maintain individual pieces of equipment and machinery such as milling machines and table saws.

Two of the most widely used organisations in schools, colleges and university workshops are:
- **DATA** (the **D**esign **a**nd **T**echnology **A**ssociation)

and

- **CLEAPSS** (**C**onsortium of **L**ocal **E**ducation **A**uthorities for the **P**rovision of **S**cience **S**ervices).

Task 10.1

Sketch or draw (IT/CAD can be used) a safety poster for a workshop you have used or are using that would include four different safety signs. The signs must cover all four areas: prohibition, mandatory, warning and no danger.

Engineering Tools and Equipment

11

In this chapter you are going to:

→ Accurately identify tools and equipment found in a workshop environment
→ Accurately state the function of tools and equipment used by Engineers
→ Select the correct tools and equipment to perform specific tasks.

To successfully achieve the objectives, you should have access to a workshop environment with enough tools and equipment to demonstrate your engineering skills (your place of learning should provide these facilities).

This chapter does not look at all the equipment in great depth but it will give you an understanding and awareness of the tools available so you can identify them and even specify equipment when a task needs to be completed. To gain further technical knowledge on how to successfully use the equipment your tutor will be able to demonstrate in the correct environment far greater technical information.

This chapter will cover the following areas of the WJEC specification:

Unit 2 LO2 Be able to plan engineering production	
AC2.1 Identify resources required	Resources: materials; equipment; tools; time
Unit 2 LO3 Be able to use engineering equipment	
AC3.1 Use tools in production of engineering products	Tools: hand tools; lathe tools; turning tools; portable power tools
AC3.2 Use equipment in production of engineering products	Equipment: centre lathes; drilling machines; milling machines; portable power tool equipment; multimeters; UV PCB lightbox; PCB tank
Unit 2 LO4 Be able to use engineering processes	
AC4.1 Use engineering processes in production of engineered products	Materials: metals; non-metals, e.g. wood, plastics Engineering processes: marking-out; cutting; finishing; preparing; shaping; drilling; turning; brazing; joining; filing; soldering

Key term

PCB: printed circuit board. A circuit board that has been made using computer aided manufacture and is in fact 'printed'.

Introduction

The technical knowledge you will gain from this and Chapter 12 is knowledge that you will be expected to demonstrate in a workshop environment for Unit 2.

You will be given a series of orthographic drawings (working drawings) and you will be expected to produce the product shown in the drawings using the knowledge you gain from this and the following chapter as well as demonstrating knowledge of the equipment and processes in the workshop.

You will also be expected to demonstrate your knowledge and application of health and safety in a working environment that your tutor will assess and grade as part of the ongoing Unit 2 process.

In Chapter 13 you will look at what work needs to be submitted to attain a positive grade in Unit 2 and examples of different formats that could be submitted to satisfy the Assessment Criteria (AC).

Machine tools

The centre lathe

The centre lathe is a machine used to manufacture mainly cylindrical products/objects. It can use different 'sections' of metal (e.g. a hexagonal section, square section, etc.) to produce shapes such as cubes, but these operations tend to be more technically advanced than just creating a cylindrical object. Centre lathes are operated both manually (in workshops) and by way of CNC in industry. Think about how many items/parts there are in the world that are cylindrical. Many different materials can be used on a centre lathe such as metals and plastics. Below is an image of a centre lathe that you would typically find in a manual workshop.

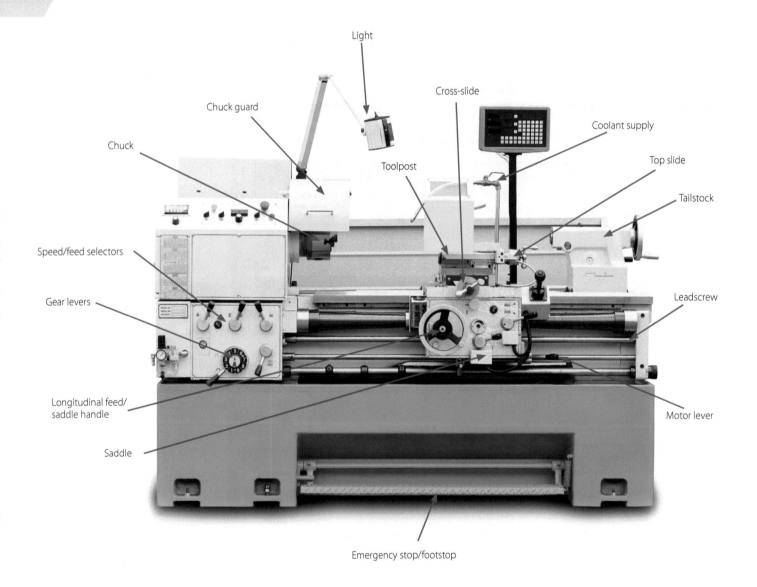

↑ A centre lathe.

Common phrases

TURNING		Reducing the **diameter** of a cylindrical object.
FACING OFF		Ensuring the **end** of a cylindrical object is flat (perpendicular to its sides).
PARTING OFF		**Cutting** the workpiece to a specific length with a specific cutting tool (parting tool).
TAPER TURNING	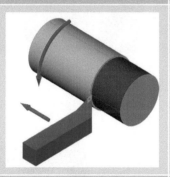	Creating a **taper** down the length of the workpiece.
KNURLING		Creating a **textured surface** on your workpiece.
GROOVING/ FACE GROOVING		Creating a **groove** on the **external diameter** or **face**.
BORING		Enlarging an existing hole in a workpiece using cutting tools or a 'boring bar'.

The vertical miller

A manual milling machine (vertical miller) is used to shape materials such as metals and plastics. Most products manufactured with milling today use CNC millers to shape different materials (e.g. aluminium phone carcasses). Milling can be very accurate when done correctly as the cutting tools are changeable and you can also mill using small diameter cutting tools for greater levels of detail.

Common phrases

FACE MILLING	Using the **bottom** of the cutting tool to 'shear' away material.
PERIPHIRAL MILLING	Using the **side** of the cutting tool to 'shear' away material.

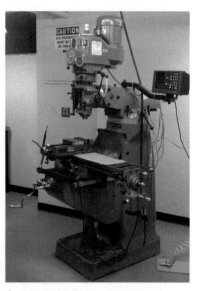

⬆ *A vertical miller.*

⬆ *Milling.*

⬆ *Milled aluminium.*

Machine drills

Machine drills are drills that are fixed in one place. Unlike hand drills (e.g. cordless), machine drills can be very accurate as the workpiece can be clamped down or held in a machine vice and the rotating drill bit is lowered using the feed lever.

There is the bench drill, a smaller type of machine drill that is bolted to a desk/bench, and there is the pillar drill that is larger and stands on the workshop floor. The larger pillar drill is more powerful and can therefore be used to drill larger diameter holes.

All machine drills have a changeable belt system that allows the user to speed-up or slow down the speed of the drill bit, depending on what material is being drilled and what diameter drill bit is being used.

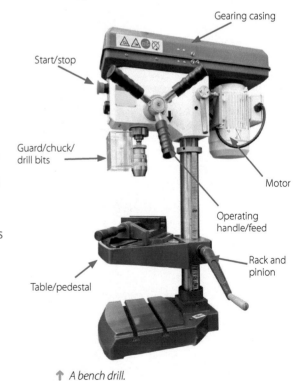

Gearing casing

Start/stop

Guard/chuck/ drill bits

Motor

Operating handle/feed

Rack and pinion

Table/pedestal

⬆ *A bench drill.*

↑ *Drilling.*

↑ *Drill bits.*

Hand drills

Hand drills are drills that are held by hand. There are many types of hand drills available (micro-drills, brace drills, egg-beater style, etc.) but the following are examples of powered hand drills.

Corded/cordless hammer drill

Corded/cordless hammer drills are very common drills that can be purchased from DIY stores. Most households have a version of this type of drill and they are used mainly to drill into masonry (walls) to hang shelves and curtain rails. These drills come with a 'hammer' setting that moves the drill bit in and out while it rotates, to create a chiselling action, making it easier to drill through masonry. They also have torque settings and can have many different features such as keyless chucks (see the following section for more on keyless chucks), LED lights, magnets, variable power/speed settings and front handles for big holes (mainly SDS drills). The cordless drills are far more popular now as they can be just as powerful as corded drills but without the problem of setting-up extension cables wherever you are working.

Key term

LED: light-emitting dioode; an electrical component that gives off or emits light.

Top tip

Depending on what drill bit you are using, you can use a hand drill to drill into most resistant materials.

Keyless chuck

Torque/hammer setting

Variable speed trigger

Forward/reverse setting

Grip

Rechargeable battery

→ *A cordless hammer drill.*

Chucks

Chucks are the part of a drill/machine that holds the drill bit (cutting tool). Chucks are also used in **centre lathes** to hold the workpiece and drill bit in the tailstock.

Corded drills (as well as machine drills) mainly have chucks that need to be used with a chuck key. The chuck key is used to loosen/tighten the jaws of the chuck and can be tightened up to a high torque setting for bigger diameter drill bits.

On cordless drills you will mainly see keyless chucks that can be tightened by hand. These types of chucks rely on a strong grip from the user to ensure a high torque fit for a drill bit. The cordless chucks have the added advantage of not having an extra part/component to carry or use and there is no danger of not being able to use the drill because of a lost chuck key.

Most chucks are three-jaw ones that are **self-centring** when using round or hexagonal sections. However, chucks on centre lathes can be changed to a four-jaw chuck for square/ octagonal sections. These four-jaw chucks need to be centred manually.

↑ *Keyless chuck.*

↑ *Jacobs chuck.*

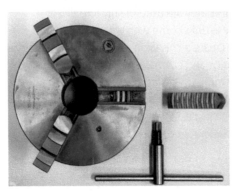

↑ *Three-jaw centre lathe chuck and key, with one jaw inverted to show the teeth.*

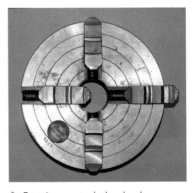

↑ *Four-jaw centre lathe chuck.*

Drill bits

Drill bits are the 'cutting tools' that are placed into the 'chuck' of a drill and rotate to cut a hole into a resistant material such as wood, metal, plastic or masonry. There are many different types of drill bits available that are used for a variety of different jobs. Below are some common drill bits that you might find at home or in a workshop.

- **Twist drill bits** are one of the most common drill bits around. They can drill most materials including metals, plastics and wood (although there are other specialist wood bits) but are no use when drilling into masonry. They are made from HSS, which is more resistant to heat than high-carbon steel and therefore does not wear as much. They are often dark grey/black in colour.

↑ *Twist drill bits.*

- **Masonry bits** are very common in households and are used for drilling into bricks/walls/masonry. They have a **chiselled** tip that is often made from tungsten carbide (a very HARD material) that is joined to the steel shaft. The chiselled tip helps to chisel away the masonry when using the 'hammer' setting on the drill. They are often shiny silver in colour.
- **Flat bits** are normally found in woodworking workshops and are used to drill larger diameter holes in wood. They leave rough edges and should only be used with a higher power drill due to the friction involved. Not to be used on metals. Used to drill fairly large holes in wood boards and sections.
- **Forstner bits** are generally used to cut blind holes (where you do not drill all the way through the wood) that are useful for hinges on different types of furniture with doors (e.g. kitchen cupboards). They come in larger diameters and should only be used with timbers (man-made and natural) or some plastics.

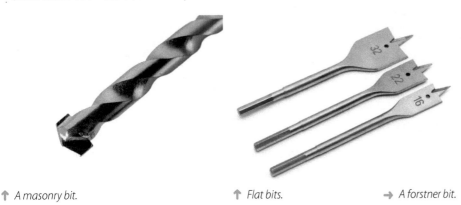 

↑ *A masonry bit.* ↑ *Flat bits.* → *A forstner bit.*

Hand tools

Engineers try-square

The engineers try-square is a tool for scribing perpendicular lines (90°) on a section of material. The stock is placed alongside the workpiece section and the blade rests on the workpiece section at 90°. The engineers try-square is also easy to slide up and down the section.

Key term

Scribe: mark-out.

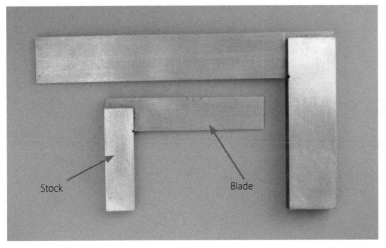

Stock Blade

↑ *Engineers try-square (metalwork).*

↑ *Try-square (woodwork).*

Scriber

A scriber is a hand tool used for marking-out areas ready for machining/cutting/drilling, etc. on workpieces made from metal. The scriber is made from high-carbon steel and is hardened to make sure it can score the surface of the metal. Marking/Engineers Blue is a liquid that can be painted onto a surface that can be scribed through to create a thin line.

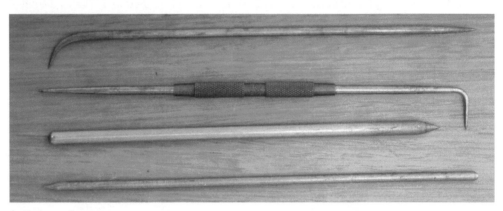

↑ *Various scribers.*

Surface gauge

A surface gauge is a scriber attached to an adjustable stand that can also be magnetised. The surface gauge can move around a flat surface to scribe horizontal lines very accurately as well as checking the accuracy of flat surfaces from processes such as milling/planing.

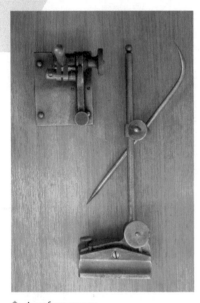

↑ *A surface gauge.*

Dividers

Dividers work very similarly to compasses. Instead of having a pencil at one end though, the dividers have scribers at both ends. This enables you to scribe circles onto metallic surfaces.

↑ *Dividers.*

V-block

A V-block is essentially a jig used to hold round or cylindrical sections of metal or plastic, while they are being marked out (scribed), drilled, or any other relevant operation. They should have clamps attached that can screw down to hold the workpiece in place while it is scribed/worked on. The V forms a 90° angle. They also come in pairs when purchased, to be used for longer sections of metal.

⬆ *A V-block.*

Centre punch

A centre punch is used for marking-out the centre of a hole in readiness for drilling. The centre punch also creates a small crater that allows the drill bit to sit-in and bite as opposed to skating around the surface and possibly drilling in the wrong place. Centre punches are generally made from steel with a hardened tip.

⬆ *A centre punch.*

Ball-pein hammer

A ball-pein hammer can be used to shape metal. It is also used for traditional processes, such as striking a centre punch, by using the flat surface. The rounded part of the head can be used to shape sheet metals or to shape the heads of rivets. Most shaping is now completed with industrial machines but the ball-pein hammer is good for small jobs in a workshop environment. Ball-pein hammers have to be tough, so are made from forged high-carbon steel (heat treated).

⬆ *A ball-pein hammer.*

Tin snips/hand shears

Tin snips are used to cut sheet metals by hand. There are larger guillotine-style shears for larger gauge (thicker) sheet metal but are cumbersome and generally fixed in one place. Tin snips are easy and quick to use by comparison but are limited for how thick the sheet material is.

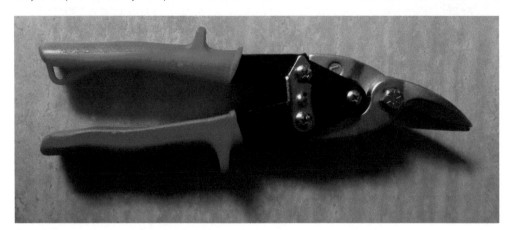

⬆ A tin snip.

Hacksaws

There are two main types of hacksaw: the standard hacksaw and the junior hacksaw. The junior hacksaw is a smaller version of the hacksaw and is sometimes easier to use on smaller jobs or in smaller spaces (plumbers sometimes use junior hacksaws to cut copper pipes because of the tight spaces they work in). The hacksaw is mainly used to cut different types of metals but can also be used on some plastics.

The teeth on a hacksaw are very small and hard, ideal for cutting harder materials such as metals. You can buy different blades with varying TPI that are used for cutting different types and thicknesses of metals but are generally sold in DIY stores, with an average of 24 TPI for most all-round jobs.

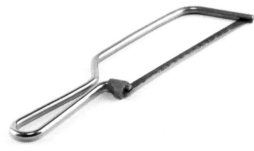

⬆ Standard hacksaw (top) and a junior hacksaw (above).

Teeth per inch (TPI)	Material/job
14TPI	Thicker metals, softer metals
18TPI	Thick to average metals
24TPI	Average metals/ general use
32TPI	Thinner metals, harder metals

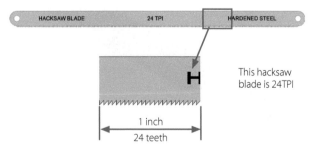

This hacksaw blade is 24TPI

Hand files

Hand files are mainly used to smooth rough surfaces on metallic objects. They can be used on certain plastics (some plastics clog-up the file) but should not be used on woods (there are other tools, such as rasps, that perform the same job for woods). Just like sandpaper, they use an abrasive surface to smooth the material and come in different grades (e.g. rough or smooth). When working on a workpiece you should start with a 'rough' grade file and eventually finish with a 'smooth' grade file.

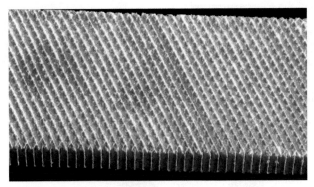

↑ *Detail from a rough grade file.*

There are many differently profiled/shaped files that are used for various jobs.

Flat files

The flat file is the most common file and is used to smooth workpieces flat. It also has a 'safe edge' that is smooth so only one surface of an internal corner can be filed.

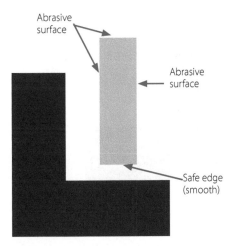

Semi-flat/half-round files

The semi-flat or half-round file is used on interior curves.

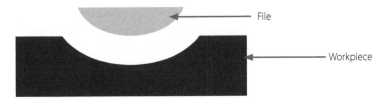

Round files

The round file is used on the interior of drilled holes. It is good for removing burrs.

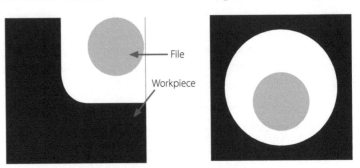

Triangular files

A triangular file can fit into very tight internal corners. It is also useful for starting a 'groove' on a flat surface.

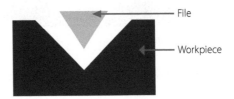

Square files

A square file is good for filing internal corners (both edges) and can file square grooves into the surface, as it is quite thin in profile.

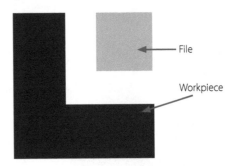

Tap and die

Tap and die sets are used to create threads on or in workpieces. With a tap and die set you can create nuts and bolts from different types of metals. Tap and die sets can also be used to clean-up existing threads by the process known as chasing but does leave the thread a little looser.

The tap is used to create/cut **INTERNAL** threads. When a hole has been drilled you can use the tap to cut a thread on the walls of the hole by effectively **screwing** the tap into the drilled hole. Taps use a tap wrench that is adjustable for different sizes. There are different shaped taps available. Tapered taps are used to make it easier to start the cutting action.

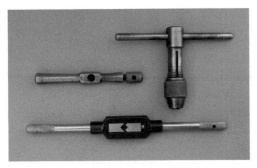

↑ Tap wrenches used for creating internal threads.

↑ Example of an internal thread that has been created by a tap.

↑ A die, die wrench and taps used for cutting external threads.

↑ This pipe shows an example of an external thread. You can see how the external thread would go into the internal thread.

The die is used to create/cut **EXTERNAL** threads. On a piece of round section metal, you can fit the die over the end and, with a rotational motion, slowly cut a thread on the exterior surface of the round metal section. Dies can be slightly opened or closed by using the screws on the **die holder/wrench**. This helps with starting the screwing and allows you to adjust the diameter setting to fit a round metal section.

To create an M8 thread you would typically need to drill a 7mm hole to allow enough material for the thread to be cut (drilled hole sizes differ depending on what size thread you need to cut, so it is always advisable to check a chart). Following is a rough guide to help you.

Tap size	M3	M4	M5	M6	M7	M8
Drill size	2.5mm	3.3mm	4.2mm	5mm	6mm	6.75mm

Top tip

You will often see on the taps and dies information explaining the sizing:
• M8 = metric 8mm.

Callipers

Callipers are used for measuring various dimensions of a workpiece. Callipers can be 'set' to a specific measurement and then used to check the measurements while work on the workpiece is in progress. This saves time using steel rulers and having to read the measurements numerous times to check accuracy, as well as offering consistency.

INTERNAL diameter measurements   EXTERNAL diameter measurements   From EDGE measurements

↑ Callipers.

Internal callipers

Internal callipers are used to check the dimensions of an internal area while it is being worked on (mainly a hole). A good example would be checking the internal diameter of a hole while you are boring-out a pre-drilled hole on a centre lathe.

External callipers

External callipers are used to check the dimensions of an external area while it is being worked on. A good example would be checking the external diameter of a workpiece that you are 'turning down' on a centre lathe.

⬆ *Internal callipers.*

⬆ *External callipers.*

Odd-leg callipers/Jenny callipers

Odd-leg callipers, also known as Jenny callipers, are used to scribe straight/parallel lines to an edge of a workpiece or section. They can also be used on circular sections of materials.

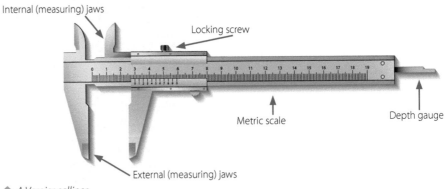

This set has a notched edge to make it easier to 'run' down an edge of a workpiece

⬆ *Two different types of odd-leg callipers.*

Vernier callipers

Vernier callipers are useful measuring tools that allow the user to measure **external diameters**, **internal diameters** and **depths** (that can also be used to measure flatter/squarer sections of workpieces). They are very accurate and can measure to 100th of a millimetre (e.g. 24.72mm).

Internal (measuring) jaws

Locking screw

Metric scale

Depth gauge

External (measuring) jaws

⬆ *A Vernier calliper.*

Micrometers

The micrometer is another useful measuring device that allows for very accurate measurements. It is mainly used to measure external diameters and external dimensions for things such as metal sections.

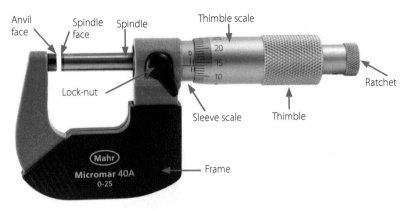

↑ *A micrometer.*

Multimeters

A multimeter is an electronic device that is used to measure the current (A – amps), voltage (V – volts) and resistance (Ω – OHMS) of a system or circuit. Some have a digital readout, others an analogue (dial) readout. They are used as a troubleshooting device to find faults within circuits and systems, as well as testing circuits or systems when carrying out maintenance. You can even test the power levels left in batteries.

↑ *A multimeter with a digital readout.*

UV PCB lightboxes and PCB tanks

Some workshops will have the equipment needed to manufacture their own printed circuit boards (PCBs). Some Engineers want to create their own equipment that may sometimes require simple circuits in order to run. With the correct knowledge, most Engineers can design their own circuits that perform simple functions. To create the simple circuits, Engineers will have to go through the process of making their own circuit boards.

To make a simple circuit board you will need to:

1. Design a simple circuit

2. Cut to size some copper-plated photoresist board.

3. Place a printed image of your circuit onto a thin plastic sheet (masking the circuit area on the copper-plating).

4. Expose the photoresist board to ultra-violet (UV) light (using a UV lightbox).

5. Place your photoresist board into developing solution.

6. Place your photoresist board into a PCB etching tank (filled with etching fluid) where the unwanted copper will dissolve away leaving only the copper tracks for the circuit you designed.

7. Start to populate your PCB with the needed components.

↑ *Example of the type of etching tank that can be found in some workshops.*

Buffing/polishing machines

The buffing machine (sometimes known as a polishing machine) is used to put a final polish/finish on a workpiece. It is mainly used to bring metals up to a highly polished finish but can also be used on some plastics. Before using this machine, your metal workpiece should first have any scratches or vice marks removed with fine files and wet and dry paper to present a smooth surface for the buffing machine. The polishing 'parts' of the machine are called 'mops'. These are natural fabric 'discs' that are stitched together with many layers. When they are spun at high speeds (up to 3,000RPM) they present a rigid surface that can be effective at polishing metals when they are pressed against them.

Key term

Wet and dry paper: abrasive sheets of paper (similar to sandpaper) for use mainly with metallic surfaces. It can be used either dry or wet. When wetted the moisture acts as a lubricant and removes the particles more quickly than when dry, creating a smoother surface, faster. Like abrasive tools and equipment (files, grinding wheel, sandpaper), wet and dry comes in different grades (grit sizes) for either a rougher or smoother finish. The lower the number/grit size of the wet and dry paper (e.g. 40 grit) the coarser the particles are on the paper and the rougher the finish, whereas the higher the grit size (e.g. 1,000 grit) the finer the particles are on the paper and the smoother the finish.

Top tip

Polish should be applied to the mops to ensure a better finish.

↑ *A buffing machine.*

Task 11.1

Copy the second and third columns of the following table into your notebook and complete them using your knowledge of tools and equipment about the image in the first column.

Tool or equipment	Name	Use/function

Task 11.1 *continued*

Tool or equipment	Name	Use/function

12

Engineering Processes

In this chapter you are going to:
➔ Learn how to correctly identify processes needed to complete a project
➔ Understand how to join (fabricate) materials effectively
➔ Understand how plastics can be formed with different moulding processes.

To successfully achieve the objectives, you should have access to a workshop environment with enough available fabricating (making) processes to demonstrate your engineering skills (your school or college should provide these facilities).

This chapter will cover the following areas of the WJEC specification:

Unit 2 LO4 Be able to use engineering processes	
AC4.1 Use engineering processes in production of engineered products	Materials: metals; non-metals, e.g. wood, plastics engineering processes Marking-out: brazing; joining; filing; soldering

Introduction

In the previous chapter, how different tools and equipment are used, the technical information needed to identify them and what they can be used for (processes) were discussed. In this chapter the focus will be more on the industrial processes that Engineers should know about so they can specify what process might be needed when undertaking a project. Engineering processes are ways of fabricating (putting together and shaping) different materials and this chapter will show you how different processes can be used to form different materials into products or parts of products.

Joining materials

Key term

Fabricate: manufacture something from different parts.

Engineers need to understand how to **join** materials. New products are mainly made-up or fabricated from different parts that have to be joined in some way. Knowing how to join the different parts is a much-needed skill.

You have probably joined or fabricated materials already, using different materials and components to join. However, there are many different methods of joining, such as using adhesives, tapes, screws, nuts and bolts, joints, hinges, welding, soldering, etc.

The joining of parts/materials is generally broken-down into two parts:
• permanent joints/fixings
• temporary joints/fixings.

Permanent joints/fixings

Permanent joints are exactly that … permanent. They are designed to not come apart during the product's life. They include:
• soldering
• brazing
• welding
• rivets
• adhesives.

Temporary joints/fixings

Temporary joints *can* last a long time but are essentially designed to not last and do come apart eventually. They include:

- adhesives
- tape
- screws
- nuts and bolts
- knock-down fittings (unusual components used to make flat-pack furniture).

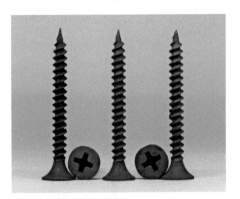  

↑ *From left to right: Phillips flathead self-tapping screws; a nut, washer and bolt; a locking cam bolt.*

The following series of engineering processes look at permanently joining materials together. This is also known as 'fabricating' products.

Soldering

Soldering is a process that is used to join metal pieces together. It is most commonly known for joining components to circuits boards or its use in the plumbing industry.

Solder is an alloy of two different metals that, when combined, create a soft metal with a very low melting temperature. Solder was usually made-up from **tin** and **lead**; however, due to restrictions on the use of lead in consumer products a lot of solder is now made from tin, copper, zinc or silver (industrial-use solder uses silver).

Due to its low melting temperature, solder can be heated and melted to form around other metals without the other metals melting. This acts as a kind of metal 'glue'.

Top tip

Flux is used on the joints you want to solder to clean the surface of the joints ready for the solder to have good contact and to allow the solder to flow into the areas that have had flux applied.

↑ *An example of the soldering process and how a component (resistor) can be permanently joined to the circuit board.*

Labels: Soldering iron, Solder, Soldered component to circuit board, Circuit board, Component

Key term

Resistor: an electrical component that can be used in a circuit to reduce/slow down the current in it.

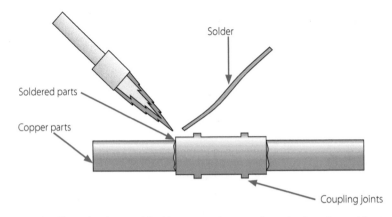

↑ *An example of how plumbers could solder copper pipes together using heat from a blowtorch.*

Brazing

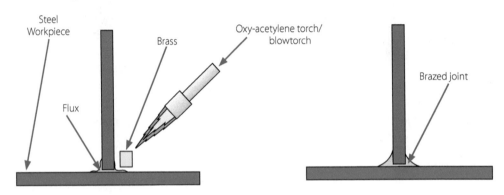

↑ *An example of the brazing process where two pieces of steel are permanently joined.*

Brazing is a process that can be used to join different metals together. It is good for joining steel (mild/stainless) to other metals as well as steel-to-steel. Brazing uses brass as the sacrifice metal to melt and join the other metals. Brass is a lot stronger than solder and therefore creates a much stronger joint. Brass also has a higher melting temperature than solder (but still lower that steel) and needs a higher temperature flame. Quite often an oxy-acetylene torch is used because of the high temperatures it produces (approximately 3,500° Celsius).

As with soldering, flux is used to clean the area to be joined and the melted brass then flows where the flux has been applied using capillary action.

You can also use 'filler rods' that are consumable metal rods made from brass instead of placing small amounts of brass in the area you want to braze/join.

Key term

Capillary action: when liquid (in the case of brazing molten metal) flows through very narrow surfaces such as two touching pieces of steel.

↑ *A filler rod.*

← *Brazing copper pipes together.*

MIG welding

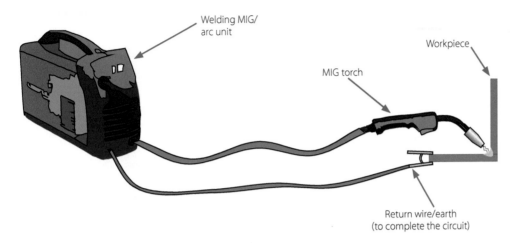

Welding MIG/
arc unit

Workpiece

MIG torch

Return wire/earth
(to complete the circuit)

↑ *An example of a MIG welding unit (left) and two pieces of steel being MIG welded (right).*

MIG welding is used to permanently join steel. It uses an electrical current to create a powerful electric arc between the steel joint you are creating to a consumable steel wire (also known as an electrode). The intense heat of the electrical arc melts the workpiece and the consumable steel wire into a molten pool where they join and create a weld. MIG welding also uses a GAS to act as a flux and clean the joint as you weld. Both the gas and the consumable wire are fed through the MIG torch. MIG welding is generally used on smaller projects with thinner materials.

ARC welding

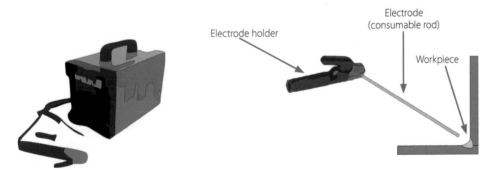

Electrode holder

Electrode
(consumable rod)

Workpiece

↑ *An example of an ARC welding unit (left) and two pieces of steel being ARC welded (right).*

ARC welding works on the same process as MIG welding, as they both use electricity to create a powerful electrical arc between a consumable rod (called an electrode) and workpiece that is hot enough to melt steel. Instead of using a wire, ARC welding uses a consumable rod, held with a rod holder. The centre part of the rod is mild steel and it is covered with flux. As you move the rod across the workpiece the rod is consumed and gets smaller. ARC welding is generally used for medium to larger sized projects with thicker materials.

> **Key term**
>
> **Electrode:** an electrical conductor that is generally used to make contact with a non-metallic part of a circuit (e.g. the use of an electrode in welding, both mig and arc, where the electrode is the 'sacrificial' metal wire or rod).

Oxy-acetylene gas welding

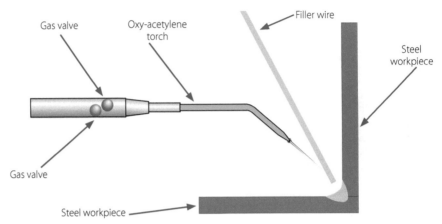

↑ *Two pieces of steel being permanently joined using the oxy-acetylene gas welding process.*

Oxy-acetylene gas welding is most commonly used to fabricate (shape/join) steel. It uses a mixture of oxygen and acetylene gases to produce an extremely hot flame (approximately 3,500° Celsius) to melt metals. The flame is so hot the steel turns into molten pools of metal. The hot steel from the filler wire and workpiece mix together and then join to form a single piece. A filler wire is also used to 'fill', which is also turned into molten metal.

Pop riveting

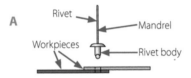

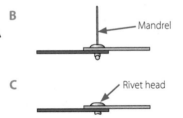

A Measure the diameter of the RIVET BODY, drill the correct sized hole in workpieces that need to be joined and insert a RIVET.

Key term

Mandrel: a cylindrical rod around which material is forged or shaped.

B Place a RIVET GUN over the mandrel and squeeze the handle as many times as needed to extract the MANDREL from the RIVET BODY.

C The workpieces should now be firmly clamped between the RIVET HEAD and the deformed end of the RIVET BODY, and the MANDREL can now be discarded.

Shaping materials

The following series of processes involves 'shaping' materials into usable forms.

When shaping metals, it is worth noting that after each 'forming' process the end product still has to be finished.

Forging

Key term

Malleable: pliable, easy to shape without breaking or cracking.

Forging is a process of joining and/or shaping metals by using force to bond workpieces together and change the form into a desired shape. The most common image people have of forging is a blacksmith using a hammer and anvil to shape heated metal. When metal is heated it becomes more malleable and is easier to shape. If the metal is hot enough it can also be bonded with the use of force (molten metal pools together to create a weld in the welding process).

1. Force is applied to two heated workpieces.

2. The heated workpieces bond.

3. After forging there is one cooled workpiece.

↑ *The process of forging.*

To heat the metals you intend to work on you would need a **forge**. Forges can be fuelled by either coal (a more traditional type) or gas (more modern). Gas forges allow the user to control the temperature more accurately and therefore allow the metals to be forged at the correct temperatures more consistently.

↑ *A blacksmith forging a product, using force and heat.*   ↑ *An older coal forge.*

Drop forging is an industrial process where the force needed to forge two workpieces together is done by a machine called a **drop forge**. The drop forge also has a die (upper and lower) in the shape of the product you are going to forge. The die is essentially a mould. A heated piece of metal, called a billet, is then placed in the die and then the upper die is dropped with force onto the billet and lower die to create the shape needed.

> **Key term**
>
> Billet: (or billet of metal) a piece of metal of a certain size that would be shaped by the forging process.

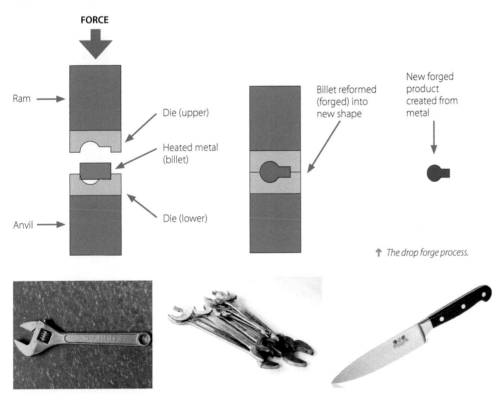

↑ *The drop forge process.*

↑ *Examples of drop-forged products.*

Casting

Casting is a process where metals are heated until molten and then poured into a mould. The metal is then left to cool and the result is a metallic object created in a desired shape. Most cast metal objects have to be FINISHED before the final object is complete. Finishing could include cutting off the sprue and runners, filing, milling, grinding, drilling and/or polishing.

Sand casting

Sand casting is an old process that is still commonly used today. The process involves taking an existing, permanent shape (pattern) and then packing SAND around that shape to create a temporary, expendable mould. The mould is then removed to create a cavity where the molten metal is poured in. Once cooled, the sand mould can be broken apart to reveal the solid metal objects in the shape of the permanent pattern. The sand can then be re-used for another mould.

The following step-by-step guide shows the main steps of sand casting and how it works.

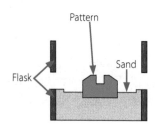

1. Sand and a permanent pattern are placed into the bottom half of a flask (sand mould holder).

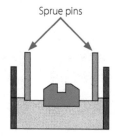

2. Sprue pins are placed to create the RISER and the top half of the flask is attached.

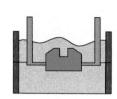

3. The rest of the flask is filled to create the mould.

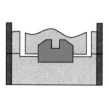

4. Sprue pins are removed.

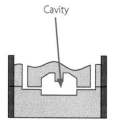

5. The pattern is removed to create the desired cavity shape.

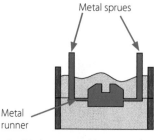

6. Molten metal is poured in to create the desired product.

7. The sprues and runners are cut away and finishing takes place to create the finished product.

↑ *Engine blocks in cars are created using sand casting.*

Moulding (plastics)

As well as metals, Engineers also use plastics to produce products. Plastics have many beneficial properties and characteristics, which make it a very popular material to work with, even though they are produced from oil, a non-renewable resource. However, bioplastics are produced from vegetable oils, corn starch and even recycled food waste. Plastics are also '**self-finishing**'. By applying different textures to the interior surface of the moulds used in shaping plastics, you can create any desired finish such as gloss, matte or textured. Following are some examples of the moulding processes for plastics.

Injection moulding

Injection moulding is a common moulding process that forces (injects) liquid plastic down a screw and into a mould. This process is accurate, is good for high-volume production, has little waste but is expensive to set up. Examples of products made in this way include PC cases and automotive parts.

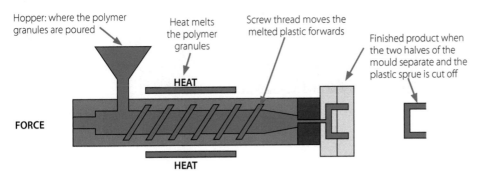

Products that have been injection moulded still need some minor finishing when they emerge from the mould, such as trimming the sprue and trimming any plastic 'bleed' from where the two halves of the mould meet.

↑ *Examples of products that can be made using injection moulding – a very accurate moulding process.*

↑ *Thermoplastic resin pellets used for injection moulding.*

Blow moulding

Blow moulding is a common moulding process that forms hot, malleable plastic to the outside of a mould. This process is inaccurate and low cost but is fast. Examples of products made in this way are drink bottles and cosmetics packaging.

The following series of four steps show how a plastic product is created using the blow moulding process.

Pre-injection moulded part is called Parison

AIR

AIR

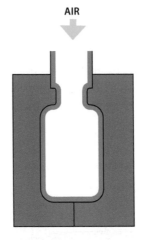
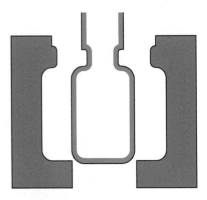

1. Mould.

2. Pre-injection moulded plastic (Parison) is clamped into the mould.

3. Air is blown into the Parison allowing it to form to the outside of the mould.

4. Mould separates leaving the finished product.

↑ *Examples of products that can be made using blow moulding.*

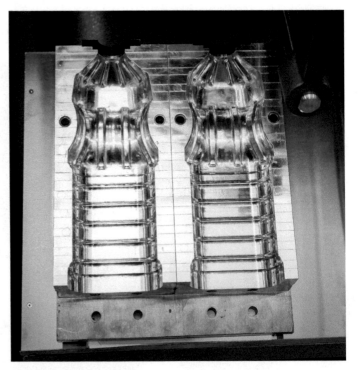

↑ *New plastic bottle moulds.*

Rotational moulding

Rotational moulding is a common moulding process that uses heat and gravity to allow plastic to form to the exterior of a mould. This process is inaccurate and low cost but is good for larger objects. Examples of products made in this way include grit bins, wheelie bins, bollards and street furniture.

The following series of four steps show how a hollow plastic product is created using the rotational moulding process.

 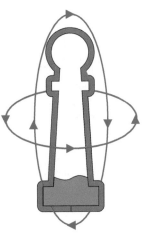  

1. Mould.

2. Polymer granules poured in, heat applied and the mould rotated.

3. Melted plastic forms to the exterior of the mould.

4. Mould separates leaving the finished product.

↑ Examples of products that can be made using rotational moulding.

← Unloading a plastic tank that has been moulded in a rotational moulding machine.

Vacuum forming

Vacuum forming is a moulding process that uses a vacuum to form sheet plastic over a pre-formed mould. This process is inaccurate, low cost and used for low-volume production. Examples of products made in this way include chocolate packaging inserts, bike helmets and cutlery trays.

The following series of four steps show how a plastic product can be created with the vacuum forming process using plastic sheets.

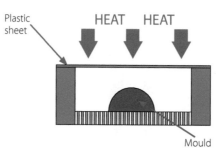

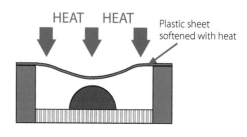

1. The mould is placed on the platen and then lowered. A thin plastic sheet (e.g. high-density polyester) is then placed on the frame above it. Heat is then applied to the plastic sheet.

2. Heat continues to be applied to the plastic until it becomes malleable.

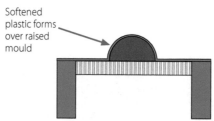

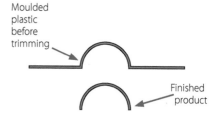

3. The platen is then raised into the plastic and a vacuum pump sucks-out all the air, ensuring the plastic forms around the mould.

4. The mould is removed from the plastic and any excess plastic is trimmed away, leaving the finished product.

Key term

Platen: the part of a vacuum forming machine that acts as a shelf with holes and can be raised or lowered.

↑ *Examples of products that can be made using vacuum forming.*

Task 12.1

Match the product to the process.

Product	Process
1. 	**A.** Drop forging
2. 	**B.** Blow moulding
3. 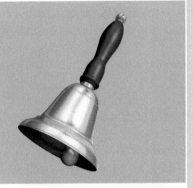	**C.** Rotational moulding
4.	**D.** Vacuum forming
5. 	**E.** Sand casting
6. 	**F.** Injection moulding

13

Unit 2 Submission

Introduction

Once you have reached this stage of the book you should have all the knowledge needed to successfully complete and submit Unit 2.

While working through Chapters 9 to 12 you will have gained a lot of technical knowledge and the skills needed to create a portfolio of relevant information and a working prototype, while demonstrating your knowledge of tools, equipment and processes.

So, what do you need to include in Unit 2?

In the **course structure** section of this book you will find a list of suggestions on content for Unit 2. Have a look at this section again to see what you will need to produce. There is also a handy portfolio guide that will help you to lay-out all the relevant information you may need to successfully complete Unit 2.

Alternatively, look at the table on page 131 to see what needs to be demonstrated, how you could demonstrate it and where to access the information needed to demonstrate your knowledge:

- The **first column** shows a list of Assessment Criteria from the specification you will have to demonstrate for Unit 2.
- The **second column** offers suggestions on HOW you can demonstrate your knowledge (each school or college will have their own interpretations of the specification and may decide to demonstrate your knowledge in different, valid ways).
- The **third column** shows you in which chapters this book has covered the relevant areas.
- The **fourth column** can be used as a checklist to see if you are happy with your knowledge or need to re-visit the chapters to further increase your knowledge.

> **Links**
>
> For information on the course structure see pages 5–6.
>
> Portfolio Guide Examples can be seen on pages 164–183.

What should I submit?

You can submit Unit 2 in any format that is easy for your school or college to work with, depending on the resources you have.

It can be submitted:
- on paper as a six–seven page portfolio, A3 or A4 (what would be best to show drawings and orthographic projections?)
- digitally
- any other format accepted by the WJEC.

Also ensure the front page of your submission clearly displays the:
- unit number (2)
- centre number and name
- candidate number.

Lastly, don't forget to keep checking the **performance bands**. Look at the work you produced and ask yourself if you are achieving the performance band you think your work deserves. Don't forget, you can always add to your work providing you do not go above the 12-hour allotted time.

The following table shows you where in this book to find the relevant text for the skills required to demonstrate knowledge of the Assessment Criteria.

CHECKLIST

Assessment Criteria	Possible ways of demonstrating	Covered in chapters:	Happy with knowledge	Re-visit chapters
AC1.1 Interpret engineering drawings	• Producing job sheets from orthographic drawings • Producing a cutting list from orthographic drawings	**1** Engineering Drawings **9** Managing and Evaluating Production		
AC1.2 Interpret engineering information	• Producing job sheets from orthographic drawings • Producing a cutting list from orthographic drawings	**1** Engineering Drawing Skills **8** Managing and Evaluating Production		
AC2.1 Identify resources required	• Producing job sheets from orthographic drawings • Producing a cutting list from orthographic drawings • Identifying machines and tools needed	**1** Engineering Drawings **3** Materials and Properties **9** Managing and Evaluating Production **11** Engineering Tools and Equipment **12** Engineering Processes		
AC2.2 Sequence required activities	• Produce a Gantt chart • Produce a sequence of tasks • Produce a learner observation record • Produce a diary of make	**3** Materials and Properties **9** Managing and Evaluating Production **11** Engineering Tools and Equipment **12** Engineering Processes		
AC3.1 Use tools in the production of engineering products	• Work safely • Use tools and equipment correctly • Produce and assemble parts to tolerance • Work with orthographic drawings • Produce a quality prototype (Tutor observed)	**3** Materials and Properties **11** Engineering Tools and Equipment **12** Engineering Processes		
AC3.2 Use equipment in production of engineering products	• Work safely • Use tools and equipment correctly • Produce and assemble parts to tolerance • Work with orthographic drawings • Produce a quality prototype (Tutor observed)	**3** Materials and Properties **11** Engineering Tools and Equipment **12** Engineering Processes		
AC4.1 Use engineering processes in production of engineered products	• Work safely • Use tools and equipment correctly • Produce and assemble parts to tolerance • Work with orthographic drawings • Produce a quality prototype (Tutor observed)	**3** Materials and Properties **9** Managing and Evaluating Production **11** Engineering Tools and Equipment **12** Engineering Processes		
AC4.2 Evaluate quality of engineered products	• Produce an evaluation (also ongoing) with evidence of reasoning	**7** Evaluating Design Ideas		

14

Effects of Engineering Achievements

In this chapter you are going to:
→ Understand some different paths of engineering
→ Recognise achievements of Engineers
→ Understand the positive effects of engineering on day-to-day life and society.

This chapter will cover the following areas of the WJEC specification:

Unit 3 LO1 Understand effects of engineering achievements	
AC1.1 Describe engineering developments	Developments: engineering (structural, mechanical, electronic); engineers involved (UK, international); key outputs; applications; technologies; materials
AC1.2 Explain effects of engineering achievements	Effects: in the home; in industry; in society

↑ *Engineers designing a jet engine.*

Introduction

In this chapter you will be looking at some of the achievements in engineering and how those success stories have had a huge impact on humanity and the world. It is quite easy to dismiss engineering as a career or job where you have to wear overalls, fix something and end up with oily hands. This stereotype of engineering is extremely inaccurate, as Engineers are the people who, throughout history, have changed the world with their innovations. For example, an Engineer is not the person who fixes an engine, but rather the person who designs and creates the engine. The jobs that require maintenance, service repair or installation tend to be populated by 'Technicians', whereas the careers that involve designing, creating and innovating tend to be populated by 'Engineers'.

However, Engineers still need the essential abilities and knowledge needed to create new equipment, machines and structures, such as understanding materials and properties, understanding and using machines tools and processes, using new technologies such as CADCAM, and 3D printing, modelling and prototyping, as well as having the fundamental skills to build their new creations from scratch. Modern Engineers are individuals with huge skill-sets and are some of the best problem solvers in the world today.

Next we will look at three areas of engineering that have had a significant impact on the world throughout history:
• Structural engineering
• Mechanical engineering
• Electronic engineering.

↑ *Technicians checking a jet engine.*

Structural engineering

Structural Engineers apply their knowledge and skills to create functioning structures such as bridges and buildings, and work as part of a team on different civic projects such as the Channel Tunnel, large dams and reservoir development. Structural Engineers can work in partnership with other professionals such as Architects to create large structures. Whereas an Architect might create the overall 'look' of the structure, a Structural Engineer would specify the materials it would be made from as well as change, amend and develop the designs to ensure the structure would function and meet not only the requirements of the client but also conforms to all the safety standards.

Following are some examples of typical structures that Structural Engineers work on.

Bridges

A **bridge** is an engineered structure that spans an obstacle such as a gorge, path or river without obstructing what is underneath the bridge. There are many different types of bridges that exist to perform lots of different functions such as carrying traffic (pedestrians, cars, trains), carrying water (aqueducts, canals) or even allowing for manoeuvrability (e.g. movable bridges for the armed forces). Engineers have been designing bridges for thousands of years to allow us to live more efficiently in our landscape.

> **Key term**
>
> Aqueduct: a bridge that carries water from one place to another.

Some of the most accomplished bridge builders of history were the Romans. The Romans discovered the strength of arches and used these shapes to create some of the most iconic ancient bridges and aqueducts in the world. Many of their engineered structures survive today.

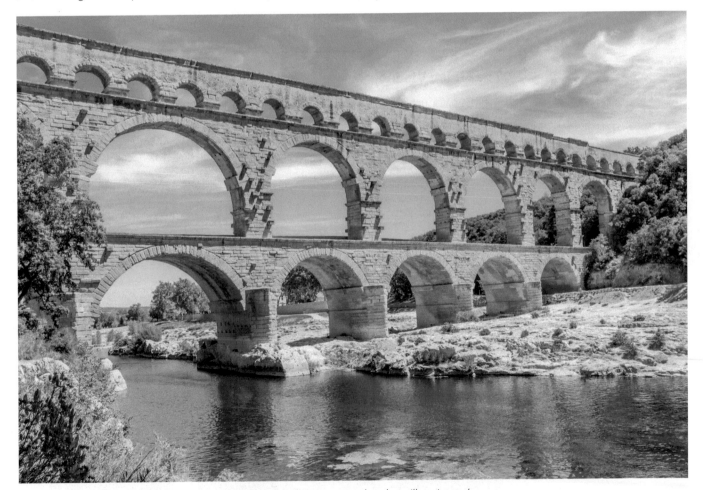

↑ *Pont du Gard, in Southern France, is an example of a well-built Roman aqueduct that still survives today.*

To understand the forces that Structural Engineers have to work with when with designing bridges, below is a diagram explaining what compression and tension are.

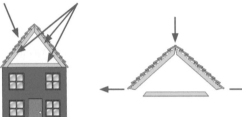

Heavy roof tiles   Three wooden joists   Weight/force of tiles

This joist stops the compressive force from pushing apart the other two joists

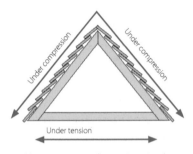

Here is a house with a cutaway of the roof. Notice how the roof has three wooden joists in a triangular shape.

The weight of the tiles puts the roof under **compression** and tries to force apart the joists making the roof collapse.

The third joist stops the two joists that are under compression from being forced apart. The third one is now being pulled apart and is under **tension**.

Key terms

Compression: being squashed.
Tension: being squeezed.

Under compression   Under compression

Under tension

Here you can see how using simple engineering you can create a roof for a house that will support lots of heavy tiles.

The Millau Viaduct

The Millau Viaduct is one of the greatest engineering achievements of modern times. It is a cable-stayed bridge that is 343 metres tall and at time of writing holds the record for being the tallest bridge in the world. It spans the valley of River Tarn near Millau in Southern France. It was also designed by a team of Structural Engineers headed-up by the British architect Sir Norman Foster.

↑ *Sir Norman Foster.*

↑ *The Millau Viaduct, Southern France.*

Following is an example of how cable-stayed bridges work. The central pylon supports the weight of the beam with a series of attached cables. The pylon is generally made from reinforced concrete, as this material performs very well under compressive forces. The cables are made from some form of steel, as steel performs very well under tension. Cable-stayed bridges often have more than one pylon to support longer spans.

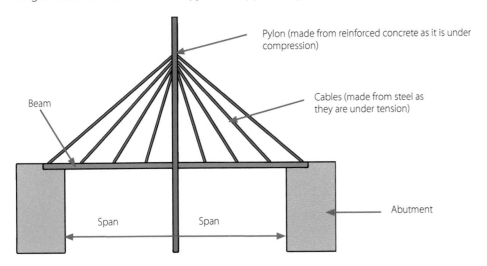

Pylon (made from reinforced concrete as it is under compression)

Cables (made from steel as they are under tension)

Beam

Abutment

Span   Span

Skyscrapers

Skyscrapers are very tall buildings with many floors and are called 'skyscrapers' as they are so tall they seem to 'scrape the sky'. They are generally built (fabricated) with a steel frame that supports all the floors and walls. The steel framework supports/bears the load of the rest of the building and all the people, furniture and equipment inside. Skyscrapers are designed and built to withstand extremely high winds, lightning and even earthquakes.

⬆ *The London skyline, showing some of the UK's tallest skyscrapers.*

When there is no more room to build outwards in a densely populated area (city) then you have to build upwards. As technology, engineering and building techniques improve, the number of floors in each skyscraper increase (at the time of writing, 160 floors was the record held by the Burj Khalifa skyscraper in Dubai, although this is set to be overtaken in 2020 by Jedddah Tower in Saudi Arabia), allowing for more space in which humans can work and live, but with only a very small footprint.

How skyscrapers are constructed

Engineers use their knowledge and understanding of materials, properties of materials and forces to construct very tall, very safe buildings such as The Shard (the UK's tallest skyscraper), which stands at 306 metres.

Following is a diagram showing how a skyscraper could be built.

A steel structure. Excellent under tension. This steel framework supports the weight of all the floors, windows, people and equipment.

Floors and exterior windows are added to the steel framework to finish the building.

A reinforced concrete foundation. Excellent under compression. This foundation supports the weight of the whole building.

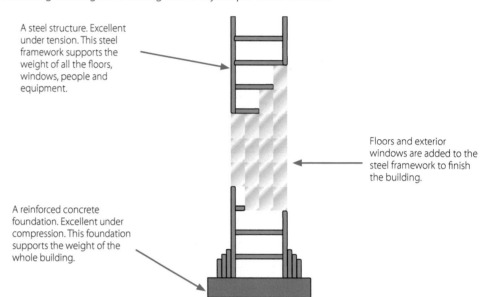

The Shard

The Shard (seen left) is one of the newest skyscrapers in the UK. It is located in Southwark, London and is 306 metres tall, making it one of the tallest buildings in the UK. It has 72 floors that can be habited and was designed by a famous Italian architect called Renzo Piano and built by a team of structural engineers from Williams Sale Partnership (WSP). The Shard is a relatively modern skyscraper that used innovative engineering techniques developed by the structural engineering team at WSP, such as top-down construction and the use of concrete in the middle of the building as well as steel. The Shard contains office spaces, restaurants, a hotel and even apartments.

Key term

Top-down construction: when constructing a building with basement floors you can complete the structure of the higher floors before excavating and constructing the lower floors.

Task 14.1

Answer the following:
1. What modern materials could you use for bridge building?
2. Explain what properties you need in a material if you are going to build a bridge and why you would need them.
3. Name a famous modern bridge.
4. Who designed the bridge you named?
5. What is the tallest skyscraper in the UK?
6. Who were the Structural Engineers that built it?
7. What materials could be used to build a skyscraper and why?

Mechanical engineering

Mechanical Engineers use their knowledge of materials, material properties, mathematics and physics to design and maintain mechanical systems. A mechanical system is a system that uses power (from a source) to perform a specific task.

For example, an early mechanical system is a water wheel (see diagram below). A flour mill, for example, uses the force of running water from a river as a power source to turn a large water wheel. The water wheel then turns a shaft that is in turn linked to a gear, which is linked to a pair of millstones. The millstones then grind grain into flour that is used to bake, for example, bread.

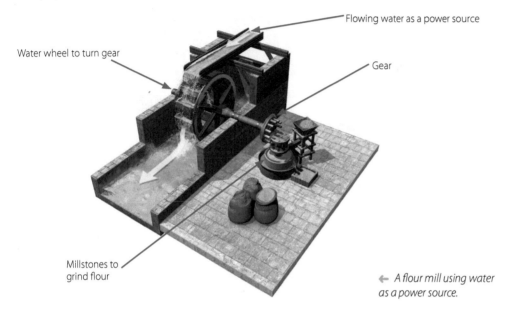

Flowing water as a power source

Water wheel to turn gear

Gear

Millstones to grind flour

← *A flour mill using water as a power source.*

There are many types of mechanical systems that make up different types of machines which have vastly improved the lives of the people and industries that use them. From simple water mills to steam-powered engines, to the latest jet propulsion and kinetic energy recovery mechanisms, Mechanical Engineers are always looking to innovate new mechanical systems for the betterment of society.

The automobile

A couple of hundred years ago most people only travelled short distances in their entire lifetime. Having to rely on horses and carts or walking, long-distance travel was not an option and daily life for the majority of people was restricted to living and working in their local vicinity. However, in today's world of advanced mechanical engineering, the average person can easily travel an average of 12,000 miles a year, because most people have and use … automobiles.

The internal combustion engine

The internal combustion engine is the power unit needed to drive most modern automobiles (with the exception of electric cars). Étienne Lenoir was the French Engineer who in 1858 invented the first commercially successful internal combustion engine. What we know as the 'modern' internal combustion engine was created in 1876 by Nikolaus Otto. Modern combustion engines are more efficient than the first combustion engines invented. The original combustion engines used to mix the fuel and air together before reaching the cylinder, while modern combustion engines inject the fuel directly into the cylinder, making it more efficient.

↑ *The Lenoir motor.*

Internal combustion engines work on the principle of using the stored power of fossil fuels to create small explosions. The power of those explosions then push a mechanical part that in turn rotates the automobile's wheels (by way of a series on linkages and gears).

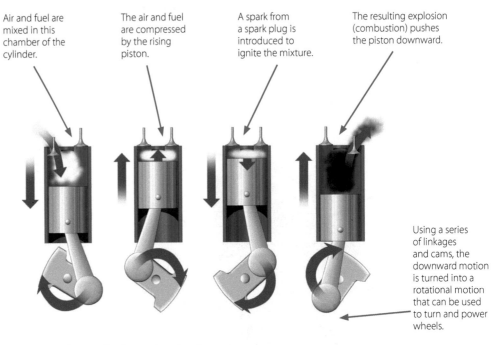

Air and fuel are mixed in this chamber of the cylinder.

The air and fuel are compressed by the rising piston.

A spark from a spark plug is introduced to ignite the mixture.

The resulting explosion (combustion) pushes the piston downward.

Using a series of linkages and cams, the downward motion is turned into a rotational motion that can be used to turn and power wheels.

→ *A cutaway image of an internal combustion engine.*

The Ford Model T

The first 'affordable' car (for working-class/middle-class American families) was the Ford Model T, which was produced from 1908 to 1927. The Ford Motor Company developed one of the first mass production assembly lines, making the product fabrication and assembly very efficient and therefore cutting down production costs.

Henry Ford (owner of the Ford Motor Company) says in his book *My Life and Work* (1922):

I will build a motor car for the great multitude. It will be large enough for the family, but small enough for the individual to run and care for. It will be constructed of the best materials, by the best men to be hired, after the simplest designs that modern engineering can devise. But it will be so low in price that no man making a good salary will be unable to own one …

He also says in the same book:

Any customer can have a car painted any color that he wants so long as it is black.

↑ *Ford Model Ts.*

Aeroplanes

Air travel has changed the world we live in and sped-up the transfer of information, people and goods.

In 1903 the Wright Brothers achieved the first successful flight and ever since then air travel has been improving as technology develops from the development of jet propulsion to unmanned drone flights.

↑ *Concorde.*

In the early 1900s it would take nearly ten days to reach North America from the UK on a ship travelling across the Atlantic Ocean. Until relatively recently, you could complete the same trip in only four hours if you could afford to fly on the supersonic passenger jet Concorde (although this is now decommissioned).

The Spitfire

The Spitfire was created as a short-range aircraft that could intercept and shoot down German bombers as well as protect outgoing bombing missions. It was a hugely successful design that (along with the Hurricane aircraft) has been credited with having been the main reason for success in the Battle of Britain. It was designed by a team of Mechanical Engineers headed-up by R. J. Mitchell, Chief Designer at Supermarine Aviation Works. One of the reasons it was so successful was its very thin wing design that allowed much higher top speeds than fighter planes of the time. Various design teams continued to work on variations of the Spitfire until the introduction of jet propulsion in the late 1930s.

⬆ *A Spitfire.*

Jet propulsion

Jet propulsion had been played around with for hundreds of years before Frank Whittle, a Royal Air Force Engineer, submitted plans to his superiors and the Patent Office in 1930 for a jet propulsion engine that could work in aeroplanes. There are different types of modern jet engine that are used in different aircraft which incorporate lots of engineering innovations. The simplest form of jet engine is the pulsejet engine, which works in the following way:

- Air is sucked in through the **air intake** where it eventually meets the **fuel injector** area and mixes with the fuel.
- A spark is then added to the mix with the **spark plug** to create a controlled explosion (combustion) where the resultant gases are expelled through the **exhaust**, which creates thrust.
- Thrust is the force that propels the aircraft through the skies.

Following is a diagram showing how jet propulsion works with a pulsejet engine.

⬆ *The statue of Frank Whittle in Coventry.*

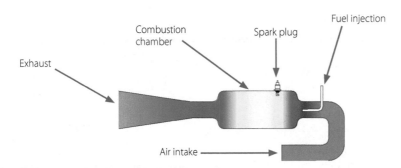
⬆ *A pulsejet engine.*

Task 14.2

Answer the following questions:
1. What modern materials could you use for an engine block in a car?
2. Explain what properties you would need in a material if you are going to build a car engine block and why you would need them.
3. Who designed the first modern internal combustion engine?
4. What was the first mass-produced car?
5. Who designed the WWII Spitfire?
6. What helped to make the Spitfire one of the fastest fighter planes of the time?
7. What force created by jets do modern aeroplanes need to fly through the sky?

Key term

Leyden jar: also known as a Leiden jar. A simple, glass jar with a foil-lined interior and exterior, used to store an electric charge – much like a battery.

↑ *Pieter van Musschenbroek.*

Electronic engineering

From early experiments trying to harness the power of lightening, to creating storable electricity with Leyden jars by Dutch scientist Pieter van Musschenbroek of Leiden (Leyden) in 1745–1746, scientists have understood the potential usefulfulness of electricity and the power it supplies. In the modern world, scientists and Engineers have developed equipment and tools that use electricity as a power source, and the creation of electronic equipment in the 20th century has changed the face of the planet and the way humans live their everyday lives. From mining new mineral elements and metals, to global communication, the introduction of processors (microchips), computers and the internet has changed the way society functions. Think of all the electronic equipment you interact with every day and then imagine a world where Engineers have not yet discovered electricity as a power source.

↑ *Electricity being used in Central Europe and the UK at night.*

Modern electronic engineering deals with the ability to design and develop electronic systems that can be controlled, and are designed to perform specific tasks. Following is an example of a basic system:

Here you can see an electronic system would have an **input** and an **output**. The input would be where you control the system and the output would be the required outcome. Let's try this diagram with a hairdryer (a simple electric system):

⬆ *The desired output from a hairdryer is hot air.*

Now the **input** would be a switch. With the switch the electric system can be controlled. The desired **output** would be the hot air … exactly what is required from the 'hairdryer system'.

Electronic Engineers now deal with very complicated systems that can perform multiple tasks. Think about your mobile phone. How many tasks does it perform? Think about not just the ability to speak to friends, take photos and listen to music, but also about your phone's ability to use light in different ways, how it applies colour, tint and tone, as well as sensing the touch of your fingertips. Within the last 30 years the advancement of electronic engineering has gone from science fiction to reality due to the achievements of Electronic Engineers.

⬆ *The Huawei Smart watch.*

⬆ *Augmented reality gaming with a virtual reality headset.*

The transistor

It could be argued that the birth of 'modern' electronic engineering began with the discovery of a simple component called a **transistor**. A transistor is a semiconductor device that can be used to either boost or amplify electronic power or switch electronic signals. Prior to the invention of transistors, electronic switches and amplifiers came in the form of vacuum tubes and other large components that made any electronic device big, cumbersome and not very energy efficient.

↑ *An example of a small vacuum tube.*

Key term

Semiconductor:
- High conductivity = conductor (e.g. metals)
- Intermediate conductivity = semiconductor (e.g. silicon)
- Low conductivity = insulator (e.g. plastics).

With transistors, electronic devices gradually grew smaller and became more efficient, which allowed Electronic Engineers to create the powerful, mobile devices that are carried today. The first practical, usable transistor was created by the American physicists Bardeen, Brattain and Shockley, who went on to share the Nobel Prize in Physics in 1956.

Following is an example of how transistors work:

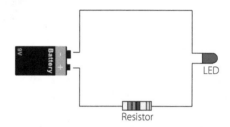

Above is a simple circuit with no transistor. To run power from the battery to the LED you need to control the main power unit (battery) with a switch. This means the circuit would have a high operational cost.

Key term

LED: Light emitting diode. A diode that emits (gives-off) light.

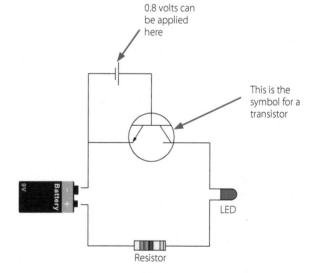

↑ *A transistor.*

Above is a simple circuit with a transistor. The LED will not light-up as the transistor is acting as an insulator and the circuit is not complete. However, when you apply a small fraction of the voltage of the main battery (e.g. 0.8 volts) the transistor turns into a conductor and allows the main power from the battery (9 volts) to run through the circuit. This means you can operate the circuit with a much lower voltage, making the system cheaper and safer.

The transistor radio

When the transistor was introduced it allowed many electronic products to become smaller and more portable. Many designers took the opportunity to develop transistor radios (radio being the main form of news and entertainment at that time) that were small, battery operated devices. In the 1960s the transistor radio was the biggest selling product in the Western world, allowing teenagers to listen to music away from home, and people to get news updates regularly wherever they were.

The iPhone

It was the late Apple CEO Steve Jobs who initially conceived the idea to have a screen that could be interacted with by way of touch for the next generation of Apple products in 2005. At the time, physical keyboards were still needed to input data into computers and phones, so designs were very limited.

From typing on physical keyboards   *to*   *interacting with a touch-screen.*

A group of Electronic Engineers were recruited to solve the problem of 'touch-screen' and investigate the viability of it working. They developed a prototype and Steve Jobs immediately looked at introducing the new technology into mobile phones. The project was called 'Project Purple 2'.

How you interact with electronic devices and the way products are designed have been changed by the team of Engineers who developed this new technology.

Task 14.3

Answer the following:
1. What electronic component changed the way electronic products were designed?
2. Describe what a 'semiconductor' is.
3. What was the best-selling consumer electronic product of the 1960s?
4. Who was responsible for changing the way we use mobile phones?
5. What was his most influential innovation?
6. What company did he work for?

15 Engineering and the Environment

In this chapter you are going to:
→ Understand how engineering affects the environment
→ See how Engineers can consider the environment in their work
→ Look at new engineering developments that have a positive effect on the environment.

This chapter will cover the following areas of the WJEC specification:

Unit 3 LO1 Understand effects of engineering achievements	
AC1.3 Explain how environmental issues affect engineering applications	Environmental issues: use; disposal; recycling; materials development; engineering processes; costs; transportation; sustainability Applications: engineering processes; engineering products

Introduction

Humans have been changing the face of the landscape since the first ancient 'Engineers' started cutting down trees to build wooden homes, right up to modern Engineers tunnelling through seabeds to create routes for rail transport. One of the biggest eras of change when Engineers had the most impact on the world and its environments was the Industrial Revolution (approximately 1760–1840). The Industrial Revolution saw a change from mainly hand-crafted products and an agrarian (farming) culture to a society where machines ruled and coal was mined to run all the newly invented steam-powered engines (used in ships, rail and industry). Factories, mass production, canals, roads and cities were built with the help of the new machines and the burning of fossil fuels, creating waste and pollution that were new parts of the world's environment.

Key term

Fossil fuels: non-renewable resources that can be burned to create energies (e.g. coal, gas and oil).

Fresh Air from the Potteries.

↑ *This image shows how the factories and machines of the Industrial Revolution emitted pollutants into the environment and changed the quality of the air that people breathed, making many of them suffer poor health.*

In the modern era we now have a much greater understanding of the impact engineering can have on the environment, and modern engineering looks to create and develop new solutions that either minimise the impact of engineering on the environment or even improve the environment through new innovations. In this chapter you will look at some of the areas and innovations Engineers are developing that have a direct positive effect on the world's environments.

Renewable energy

Prior to the development of technologies that have allowed us to harness the power of renewable energies, societies across the world only burned fossil fuels to create power. In fact, fossil fuels are still the main source of power for the world today, with oil being the resource that is predominantly used to power industrial machines and transport, and create materials such as plastics. The problem with using fossil fuels (a non-renewable resource) is that that they not only cause harm to the environment by the pollutants they create when burnt but they are also a finite resource and will eventually run-out.

> **Key term**
>
> Renewable energies: energies that can be renewed naturally such as wind, solar, geothermal and tidal.

↑ *A coal-fired power station releasing CO_2 into the atmosphere.*

Due to the huge drawbacks of using fossil fuels, Engineers have been looking at creating technologies that harness the use of renewable energy sources such as wind, solar, hydro and geothermal.

Wind power

Wind power or wind energy uses wind turbines to capture the energy in the movement of air. The turbines use propellers to turn a generator to create electrical energy that is then either stored in batteries or fed directly into the electric grid. Wind farms are dedicated areas where lots of wind turbines can be grouped together to capture the energy of air currents. Wind turbines can be expensive to maintain as the mechanical nature of the turbines means they need to be regularly maintained; however, they have a minimal impact on the environment and make good use of a renewable energy source.

↑ *Wind turbines on a hillside, capturing the energy of the movement of air.*

Solar power

Solar power is the ability to catch and convert sunlight into electricity that is either stored in batteries or fed directly into a power grid. Sunlight is captured and converted using solar panels, which are now quite a common product and can be seen adorning the rooftops of many houses. Solar panels use photovoltaic cells to convert sunlight into electricity by allowing photons to knock electrons free from atoms which then create a flow of electricity.

↑ *The largest solar plant in the world is in the Mojave Desert, USA, at 3,500 acres.*

However, to run solar panels efficiently you still need to have good air quality to ensure the sunlight can actually reach the solar panels. For example, China (one of the world's largest manufacturing economies) still burns huge amounts of fossil fuels for its industries, alongside the use of newer technologies such as solar panels. This is causing a problem for the new technologies, as the air quality is so poor in some areas that the smog and air pollution block the sunlight from reaching the solar panels because it is so thick (source: Fabienne Lang (2019, 11 July) 'China's Air Pollution is so Bad it's Blocking its Solar Panels', *Interesting Engineering*, https://interestingengineering.com/chinas-air-pollution-is-so-bad-its-blocking-its-solar-panels).

Hydropower

Hydropower involves harnessing the power of moving bodies of water (e.g. tides) to turn turbines that create electricity which is then either stored in batteries or fed directly into power grids.

↑ *A hydroelectric plant built on a river.*

↑ *Tidal energy turning turbines to create electricity.*

The use of water to power machines has been in use for centuries, with early watermills using the power of the flow of a river to turn a wheel, which in turn would turn grindstones for grinding grain. Hydropower can be obtained in several ways, from building large dams and reservoirs to create hydroelectric plants, to creating floating turbines out at sea to harness the power of the waves. A great advantage in using water to create electricity is that it is easy to predict the tides, the flow of a river or even when you open the sluice gates of a dam to allow water to flow. This predictability makes hydropower a potentially very efficient source of renewable energy.

Geothermal power

Geothermal power is the harnessing of the Earth's heat to raise the temperature of water and create steam. The steam then turns turbines which convert the steam into electricity. This is a clean and renewable resource that is currently being used by many countries in the world including Kenya, Iceland and New Zealand, which are generating over 15% of their countries' energy needs from geothermal sources. However, although geothermal energy has many benefits to the environment, it is expensive to create geothermal plants and find locations in the world where they can be built.

↑ *The power of the river flow turns the wheel – early hydropower.*

↑ *A geothermal power plant in New Zealand.*

The product lifecycle

When creating new innovations and solutions, Engineers need to understand the impact their design is going to have on the planet. Modern Engineers must ask, what will it be made from, where will I obtain the resources needed, how will it be used and what will happen to it when it comes to the end of its life? By addressing these questions, modern Engineers can produce products and solutions that may minimise their solution's impact on the world's environment. The responsibility of using the world's resources lies with the decisions made by the Engineer when starting a project and making choices that ensure there is minimal damage to the environment.

Understanding these issues has led to the development of the product lifecycle assessment. This model looks at the overall impact of the creation of a new product and allows decisions to be made at each stage that could potentially minimise the harmful impact on the environment.

End of product life
What happens to the product at the end of its life? Can any or all of it be recycled/re-used? Is it easy to dismantle and recycle? Will it go into landfill (rubbish dump)? How can this now useless product have a minimal impact on the environment?

Extracting raw materials
What and where are the materials you need for your solution? Do you have to transport them from the other side of the world? Can you source them locally? How are you going to extract them? Can you find recycled materials to use?

Using the product
How will the product be used? Have you created it to only last a limited time? Is it an optimal design that will last a long time? Does it need servicing or maintenance? Are there extra environmental costs if the customer uses it (e.g. power usage)?

PRODUCT LIFECYCLE ASSESSMENT

Refining raw materials
Do the materials you have specified need refining (e.g. crude oil into plastic)? How many refining processes are your materials going to need? Can you substitute some raw materials for recycled/re-used materials that have already been refined?

Assembling parts
How is your product going to be assembled and packaged? Will your packaging use even more material? Can it be assembled in the same place as it was manufactured? Once assembled, does it need to be transported to the customers?

Manufacturing parts
Where is your product going to be manufactured? Does it need to be transported when made? Do you need to set up a new manufacturing plant and train new staff? Can you manufacture locally?

The engineering design process can have a major impact on each of the processes listed. Engineers, therefore, have a huge responsibility in looking after the environment and need to consider design choices that would affect each of the above areas. Can you choose a better material? Use less transport? Make the product re-usable or recyclable?

Task 15.1

Following is an unfinished diagram of a product lifecycle assessment for an aluminium can of cola that you could buy from your local shop. Copy the diagram and complete each stage of the product lifecycle assessment for this product, then write a short paragraph describing the impact the aluminium can has had on the environment.

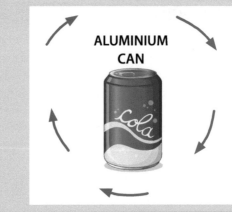

ALUMINIUM CAN

Existing and future engineering materials and processes

Engineers and scientists all around the world are constantly looking to develop new materials and engineering processes that are not only more efficient and cost effective but are also better for the environment and are sustainable in the long term. Whether it is reducing the amount of raw materials used by substituting recyclable materials or developing new and exciting ways of building and manufacturing, Engineers should now prioritise the environment when creating new solutions and incorporating new technological advancements when making decisions. Here are a few examples of some innovative practices that are currently being used or will be used in product and engineering developments.

- **Sustainable concrete**: fewer actual materials can be used to create large volumes of concrete (an excellent building material) by using items such as crushed glass, wood chips and slag to add bulk to the overall mix.
- **Pollution absorbing bricks**: these bricks cannot only be used for construction but can also act as air filters for the immediate environment around them. They can absorb fine to coarse particles and reduce air pollution.
- **Bioplastic**: instead of using chemicals from crude oil, bioplastics derive from organic matter and plants such as sugar cane, algae, corn starch and crustaceans. Apart from not using non-renewable fossil fuels, bioplastics are also 100% biodegradable, ensuring there are no harmful effects to the environment when disposed of.

> **Key term**
>
> Slag: waste material that is left when smelting or refining ore from metals.

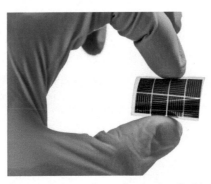

↑ *Disposable plates and cutlery made from bioplastic.*   ↑ *A small photovoltaic cell printed on flexible film.*

- **Photovoltaic surfaces**: much like solar panels, photovoltaic surfaces are an advanced use of solar power technology but, unlike solar panels, can be applied to surfaces such as glass. Large structures such as skyscrapers could be self-powered and self-sustainable if covered in a photovoltaic film.
- **Self-healing materials**: although in the early stages of development, this is a material that has the potential to 'self-heal'. By using the carbon in the atmosphere, self-healing materials could repair themselves when broken.
- **Smart factories**: there is currently 'very clever' manufacturing using computer-integrated manufacturing and robotics, but smart factories are the next level. Smart factories are able to adapt and change processes through real-time monitoring, as every aspect of the manufacturing process is managed by digital computers and robotics. This means that there would be nearly no human input in the day-to-day manufacturing of products in a smart factory. Good news in terms of efficiency; bad news in terms of employment opportunities.

↑ *An Engineer uses a tablet to monitor a robot arm in a smart factory.*

What is sustainability/sustainable engineering?

Sustainability (in terms of engineering and creating new products) means:
- creating products that are made from sustainable resources
- creating products using minimal/renewable resources during manufacture and transport
- creating products that can be recycled fully.

As an example, a team of Structural Engineers sets out to create a sustainable office building. To achieve this goal they may have to:
- source materials from a renewable resource
- use materials that are 100% recyclable
- minimise the use of any non-sustainable materials
- source materials locally
- create a building design that has a minimum impact on the environment
- ensure the building operates efficiently in terms of energy use (renewable energy sources)
- ensure there is minimal maintenance needed for the building
- ensure the building is easy to renovate and update if needed
- ensure the building is easy to dismantle, demolish and recycle at the end of its life.

Sustainable engineering not only minimises the impact of creating new products by not using 'virgin' resources but also has a positive impact on environments.

↑ The Crystal (London) is an all-electric sustainable building that uses solar power and a ground-source heat pump to generate its own energy. The building incorporates rainwater harvesting, black water treatment, solar heating and automated building management systems (source: https://www.thecrystal.org/about/).

Recycling materials

Millions of tons of materials are used every year to produce products from plastic packaging to new buildings, with a lot of the materials eventually ending up in landfill sites or being thrown into the oceans as waste instead of being re-purposed or recycled. The new goal for engineering and manufacturing is to create a material-neutral world, where all the resources we extract could be used again and again, or fully biodegrade into a non-harmful (to the environment) substance. Clearly, we are not there yet but we are getting closer as innovations and technology improve. A good example would be using bioplastics instead of plastics derived from crude oil, as bioplastic is 100% biodegradable.

However, the recycling of materials is now law, with ISO (International Standardisation Organisation) setting world standards for the percentage of materials that have to be recyclable when new products are manufactured. For example, the standardisation number ISO 15270:2008 deals with what percentage of plastics have to be recyclable for all new products, and all manufacturing companies have to conform to this international law.

Emblems and labelling for recycling and sustainability

When you use or purchase a new product you may see logos printed or stamped onto the packaging or surface, showing if the product comes from a sustainable source or if parts of it can be recycled. Here are some examples of logos that are placed on products to help the consumer make a choice.

♲	**Recycling logo (mobius loop)**	Shows if a product or part of a product can be recycled (plastic products, packaging).
FSC	**Forest Stewardship Council®**	Shows that a forest product (e.g. wood or paper) is responsibly sourced.
EU Ecolabel www.ecolabel.eu	**EU Ecolabel**	Shows if a product has conformed to European standards for sustainability (products manufactured in Europe).

Recycling plastics

Most plastics can now be recycled. However, the recycling process for plastics can be difficult and does tend to use lots of energy. Plastic also degrades every time it is recycled and can only be recycled a limited number of times. All plastic products now have to be stamped with the type of plastic it is to make it easier to recycle and reduce costs.

Recycling is an expensive process, so any well-engineered design that makes it easier to recycle will reduce the cost to society and the environment, and this is why modern Engineers work hard to create solutions and products that consumers and industry find easy to recycle.

⬆ *Most plastics are recyclable.*

Following is a plastic recycling chart that shows what the 'stamped' numbers on plastic products mean and what materials they are made from.

PET	HDPE	PVC	LDPE	PP	PS	OTHER
polyethylene terephtalate	high-density polyethylene	polyvinyl chloride	low-density polyethylene	polypropylene	polystyrene	
soft drinks bottles	milk bottles	chocolate box trays	shopping bags	toys	toys	other plastics such as:
juice containers	cleaning agents	clear plastic packaging	squeezable bottles	luggage	hard packaging items	fibreglass
cooking oil bottles	laundry detergents	bubble wrap	plastic sacks	car bumpers	CD cases	acrylic
	shampoo bottles	food packaging		plastic chairs		nylon

Task 15.2

Find three plastic products that have been stamped with a material recycling number. Use the chart on the right to identify what material they are made from.

Recycling metals

Using metals in products and construction is generally good news, as metals can be considered to be sustainable materials because they are almost 100% recyclable (apart from a small percentage of corrosion from ferrous metals) and re-usable. You can recycle metals again and again without changing their properties. This is good news for Engineers and the environment. One of the most common metals used today is steel, which is probably the most recycled metal, so that steel car you drive around in could have previously been a washing machine.

⬆ *A selection of ferrous and non-ferrous metals.*

Task 15.3

Using your new knowledge on engineering and the environment, answer the following:
1. Name three renewable energy sources.
2. Explain the term 'sustainability'.
3. What building materials could be considered 'sustainable' and why?
4. Which organisation sets standards on recycling targets for new products?
5. What recycling logo could be displayed on new products and what benefits would this bring to the consumer?

Engineering Mathematical Techniques

In this chapter you are going to:
→ Learn how to correctly identify metric measurements
→ Learn how to use simple mathematical formulae to solve engineering/mathematical problems
→ Learn how to work out the values of an electronic circuit.

This chapter will cover the following areas of the WJEC specification:

Unit 3 LO4 Be able to solve engineering problems	
AC4.1 Use mathematical techniques for solving engineering problems	Mathematical techniques: use of formulae; Ohm's law; efficiency; areas and volumes of geometric shapes; calculation; measuring; estimation; mean; units of measurement; metric; metres, millimetres; pounds, pence

Introduction

Understanding how to use mathematical techniques is a fundamental skill that all Engineers should have. Engineers are commonly faced with problems that lead to the questions, how big, how heavy or how much? When building bridges, Engineers need to know how heavy the traffic might be, which would be a deciding factor in what materials they would specify for the task; or need to know what components would be used in a circuit and therefore how much power would be needed to run the circuit. Mathematics and engineering go hand-in-hand and an Engineer would not progress very far without some knowledge of this subject.

Mathematics can sometimes be seen as a difficult subject and might deter some people from considering engineering. However, if you approach mathematics in a simple step-by-step method it can be very easy to understand. In this chapter you are going to learn some engineering mathematical techniques in easy to access step-by-step processes that will give you a greater understanding of how mathematics can be applied to simple engineering problems.

Areas

Engineers always need to understand how to work out areas, as this exercise allows them to answer the question: how big is it?

Imagine you are a Structural Engineer and need to work out how many new houses you can fit onto a newly purchased development site. The first thing you might do is find out how much ground space each house will use and would therefore need to find out the area of the footprint for each house.

The area of a shape is a measure of the 2D space that it covers. Area is measured in squares, for example square centimetres, square metres and square kilometres.

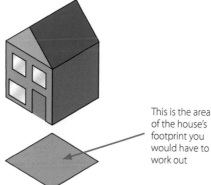

This is the area of the house's footprint you would have to work out

Area of rectangles

The following formula should be used to work out the areas of rectangles and squares:

A = Area

L = Length

W = Width

A = L × W

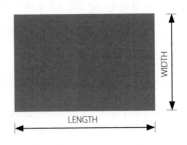

Area of parallelograms

The area of a parallelogram is the Base × Perpendicular height (B × H):

A = B × H (perpendicular)

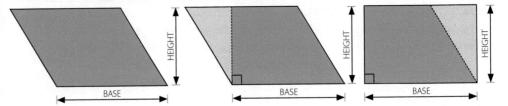

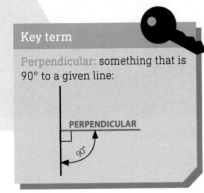

Area of triangles

If you multiply the base by the perpendicular height, you get the area of a rectangle. The area of the triangle is **half** the area of the rectangle.

So, to find the area of a triangle, multiply the base by the perpendicular height and divide by two.

The formula is:

A = (B × H)/2

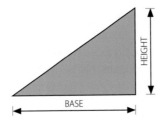

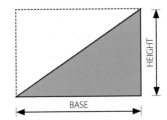

The below non-right-angle triangle uses the same principle. All you need to do is find the perpendicular line to measure the height. You will then be left with TWO right-angle triangles to work out. Then you just add them together.

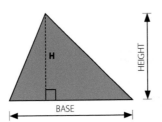

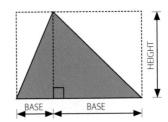

Area of circles

To find out the area of a circle you must first know what the radius and diameter are.

The diameter goes from one edge of the circle, straight through the centre point to the opposite edge.

The radius goes from the centre point to the edge of the circle.

The formula for finding the area of a circle is:

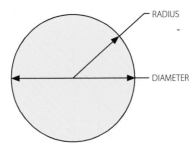

$A = \pi r^2$

A = area

$\pi = 3.14159265359$

r = radius

RADIUS

DIAMETER

So:

$A = 3.14 \times (r \times r)$

To find the area of HALF a circle is:

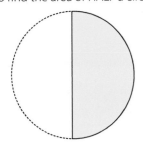

$A = 3.14 \times (r \times r) \div 2$

To find the area of QUARTER of a circle is:

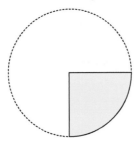

$A = 3.14 \times (r \times r) \div 4$

Area of compound shapes

A compound shape is a non-standard shape that looks complicated but is actually very simple to work out.

The shape below looks complicated and might cause you trouble when trying to work out the area in its entirety.

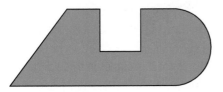

A simple solution to working out compound shapes is to break them up into very simple shapes, work out the area of each shape (see previous area examples) and then either ADD or SUBTRACT the individual areas depending on how you broke them up.

Either 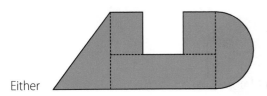 or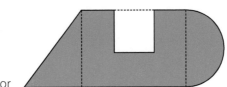

Task 16.1

Using your learnt knowledge of areas, work out the area of the following aluminium bracket. If you get stuck, just look back at how to break-up complicated shapes into simple shapes and then work out the areas for each simple shape. Show ALL your workings out.

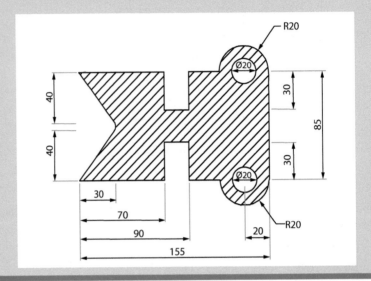

Volumes

Volumes are another mathematical technique that Engineers must learn. A good example of this is understanding the volume of steel you might need to support a bridge and all the traffic that it will have to support. Too little steel may cause the bridge to collapse and too much steel may cause the bridge to be too heavy and too costly to build.

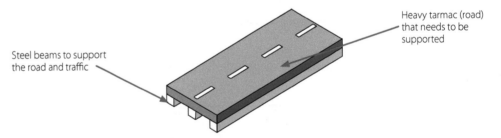

Steel beams to support the road and traffic

Heavy tarmac (road) that needs to be supported

The **volume** of a shape is all the space it occupies in three dimensions, whereas the **area** only measures the surface (dimensions).

The **volume** of a shape is measured and written as **cubed** or ^{3}.

Below is a cubic centimetre. Each of its sides measures 1cm in length. Therefore, the volume of the 3D shape is 1cm^3.

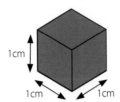

1cm

1cm   1cm

The following 3D shape has a total of eight 1cm³ cubes. So, the volume of it is 8cm³.

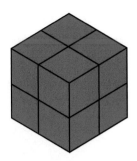

Volume of cuboids

A cuboid is a shape that has six sides that are all perpendicular to each other (90°). A good example would be a solid brick.

To find the volume of a cuboid, multiply its **length by its width by its height**. We can write this as:

$$V = L \times W \times H$$

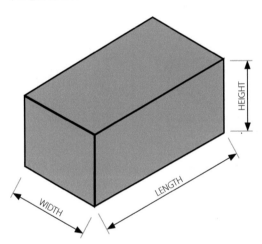

Task 16.2

How many cubes are there in the following two shapes? Write down the volume.

1.

2.

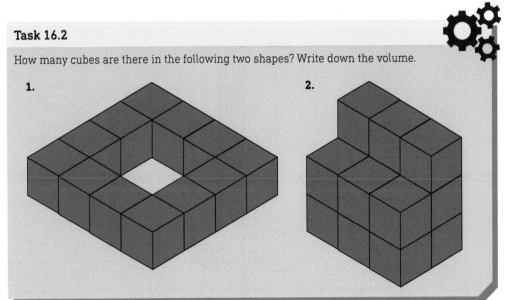

Volume of prisms

You have already learnt how to work out areas. If the area was the end of a section of mild steel (e.g. a square section mild steel) that would also be known as a **cross-section**. The length of the steel could also be known as a prism.

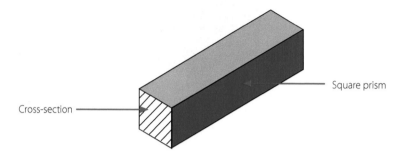

Cross-section

Square prism

This formula works for all prisms:

Volume = Area of cross-section × Length

Different types of prism

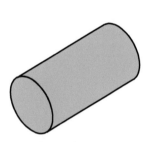

Volume of a cylinder
= Area of circle × Length

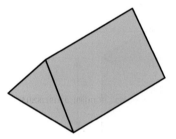

Volume of triangular prism
= Area of triangle × Length

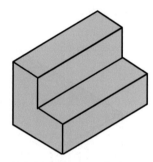

Volume of 'L'-shaped prism
= Area of 'L'-shape × Length

Task 16.3

Using your learnt knowledge of areas and volumes, work out the volume of the below steel rivet. Show ALL your workings out.

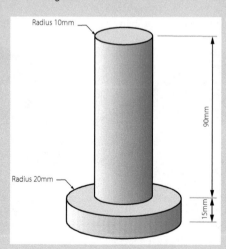

Radius 10mm

90mm

Radius 20mm

15mm

The mean (average)

The mean is the most common measure of average. If you ask someone to find the average, this is the method they are likely to use.

To find out the average from a set of data, ADD all the data (numbers) together and then DIVIDE the result by the total amount of numbers (data), for example:

Below is a set of scores taken from students' engineering tests (11 tests, 11 scores):

9, 13, 9, 11, 9, 13, 11, 9, 10, 8, 11.

Added together the total is: 113.

So, the MEAN is: 113 ÷ 11 = 10.27.

Now we know the average score for the test for that particular set is 10.27 per student.

Task 16.4

A family has bought a car. They want to know how much it is costing to run the car in fuel costs per month. They want to know the MEAN cost and they have recorded the miles travelled each month for a year.

The new car does 10MPL (miles per litre).

A litre of fuel costs £1.00.

Month	Jan.	Feb.	Mar.	Apr.	May	Jun.	Jul.	Aug.	Sep.	Oct.	Nov.	Dec.
Miles	600	723	650	760	667	544	556	700	801	599	655	745

Efficiency

All Engineers need to contend with and understand the use of energy. Sometimes it is the use of energy to produce a product; other times it could be the use of energy of a product they have designed.

Engineers deal with different types of energy such as:
- heat
- light
- kinetic
- chemical
- electrical
- sound
- gravity
- elasticity.

ENERGY IS MEASURED IN JOULES (J).

When looking at products and processes, Engineers need to understand how much energy would be needed to run a product or process, how much of that energy is used and how much of that energy is lost. This is called **EFFICIENCY**.

Look at the below diagram. You can see a simple filament lightbulb being used. An amount of electrical energy is used to run the product/process. The desired result would be 'light'. However, during the process you can see how some energy is being lost with 'heat' and, to a lesser extent, 'sound'.

If you add values to each of these, you can then work out the **efficiency** of the lightbulb.

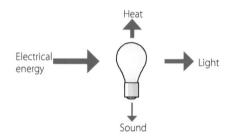

To work out the efficiency of a product/process you can use the following formula (efficiency is written as a percentage: %):

$$\text{Efficiency (\%)} = \frac{\text{Useful energy OUT (J)}}{\text{Total energy IN (J)}} (\times 100)$$

Task 16.5

What is the efficiency of the following lightbulb?

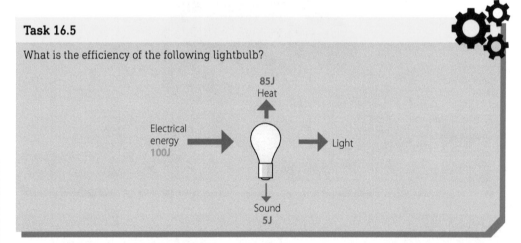

Mathematical techniques for electronics

Having a basic understanding of simple circuits should be part of an Engineer's skill set, as electronics and circuits are now used in many different products.

In this section you will need to understand about **voltage**, **current** and **resistance**.

Ohm's law

Ohm's law is a formula that is used to work out the resistance in a circuit.

Circuits are made-up from:

Voltage (the power supply of the circuit, the *PUSH*, e.g. 9-volt batteries)

Current (the amount of electricity running around the circuit)

Resistance (how the current is slowed down by encountering things in the way, e.g. wires, components)

↑ *Georg Ohm Memorial in Munich, whose work Ohm's law is named after.*

The following diagram shows how you can work out the values of each:

Voltage (V) $= \text{Current} \times \text{Resistance}$

Current (I) $= \dfrac{\text{Voltage}}{\text{Resistance}}$

Resistance (R) $= \dfrac{\text{Voltage}}{\text{Current}}$

When measuring the values of voltage, current and resistance the units used are:

Voltage – volts (V)

Current – amps (A)

Resistance – ohms (Ω)

The following is a simple 'series' circuit with four component parts.

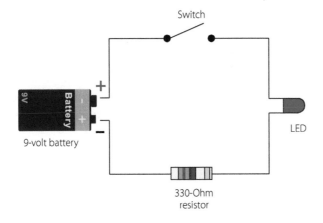

Switch

9 V Battery
9-volt battery

LED

330-Ohm resistor

Using Ohm's law, calculate the current (I) of the above circuit:

Current (I) $= \dfrac{\text{Voltage (V)}}{\text{Resistance (R)}}$ ____ amps $=$ ____ $\dfrac{\text{volts}}{\text{ohms}}$

Task 16.6

Using OHM'S LAW, solve the following problems:
1. A car has an internal cabin lightbulb to light up the interior of the car. The lightbulb has a resistance of 24Ω (R). The current running through the lightbulb is 0.5 amps (I). What voltage is needed to run the lightbulb?

2. A small keyring torch is run with a current of 0.3 amps taken from a 3-volt battery. What is the resistance of the circuit?

Resistance (R) $= \dfrac{\text{Voltage}}{\text{Current}}$

Electronic components

As you have discovered, electronic circuits are populated with different components. These components all have different jobs to do and when put together can create electronic systems that perform specific tasks. An Engineer needs to be able to identify electronic components to design circuits for electronic systems.

When designing and drawing circuits, electronic components are represented by symbols. The following symbols are examples of the components that can be used in a simple circuit for an LED light:

	The symbol for a resistor.	A resistor resists the current in a circuit and reduces the flow of current to protect other components that would be destroyed if too much current flowed through them.
	The symbol for a switch.	A switch 'opens' or 'closes' a circuit, thereby allowing the current to flow or not flow.
	The symbol for a battery.	A battery stores electrical energy and acts as a power source for part or the whole of the circuit.
	The symbol for an LED.	An LED is a component that emits light when a current flows through it.

Now let's put these circuit symbols into a circuit diagram:

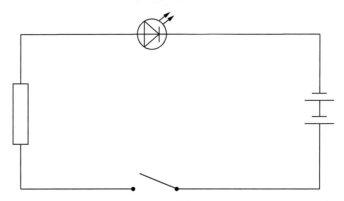

⬆ *A simple circuit diagram showing a circuit that creates light with an LED by using component symbols.*

By understanding the symbols used for components you can quickly identify what the function of a circuit is just by looking at the diagram.

Task 16.7

Following is a table that shows some very common component symbols (you already know some of them). Research the component symbols and copy and complete the chart in your notebook, making sure you write down the function of each component.

Component symbol	Component name	Component function
▭		
▭		
▱ (arrow)		
▱		
⊡		
⁄		
⊓		
⊢⊦		
⊣⊢⊦		
⊳		
⊳ (arrows)		
⊗		
⊕		
Ⓐ		
Ⓥ		

Portfolio Guide Examples

Unit 1

Derry Accessories Ltd – EXAMPLE BRIEF

Derry Accessories Ltd manufactures parts and accessories for mobile phones. The company successfully manufactures replacement mobile phone chargers, similar to the one in the photograph on the left. Derry Accessories Ltd manufactures many different chargers for different brands of mobile phones. All current models of charger include a cable which links the mobile phone to an electrical plug. The company has been successful for a number of reasons:

- There is a large range of mobile phones on the market, each range needing its own type of charger.
- Mobile phone manufacturers are continually developing new models and Derry Accessories Ltd ensures it has a replacement charger available.
- New mobile phones have a shorter battery life and need regular charging and many people need more than one charger so their phones can be charged at home or elsewhere.

However, these issues also have a negative impact on the company, as each charger requires the design team to develop a new specification which uses slightly different materials and components. This increases design, manufacturing and warehousing costs. There are also environmental impacts, as each charger uses materials that are not biodegradable. These would, however, be recyclable and broken down into their component parts as part of the Waste Electrical and Electronic Equipment (WEEE) directive.

Feedback from retailers and mobile phone manufacturers indicates that consumers have a number of concerns about replacement chargers. These include:

- Mobile phones being charged in homes and offices are generally unsightly.
- Many households charge two or more phones at any time.
- In businesses, employees use facilities to charge their phones, which can look unprofessional.
- Existing chargers do not complement the style of modern phones.
- Over time the cable loses elasticity becoming unsightly in appearance.

Design brief

Derry Accessories Ltd is looking to design a new 'generic' type of mobile phone charger. The company is prepared to consider any options for charging mobile phones but is eager to ensure that all phone functions are accessible when charging. Derry Accessories Ltd views the development of this new model as an opportunity to promote the company and wants to see their logo on all future products.

Your task is to design a new generic mobile phone charger.

Copy the design brief into your notebook and highlight the key features with a highlighter … also add an explanation as to what each of the highlighted areas means.

Product analysis/identifying functions of engineered products

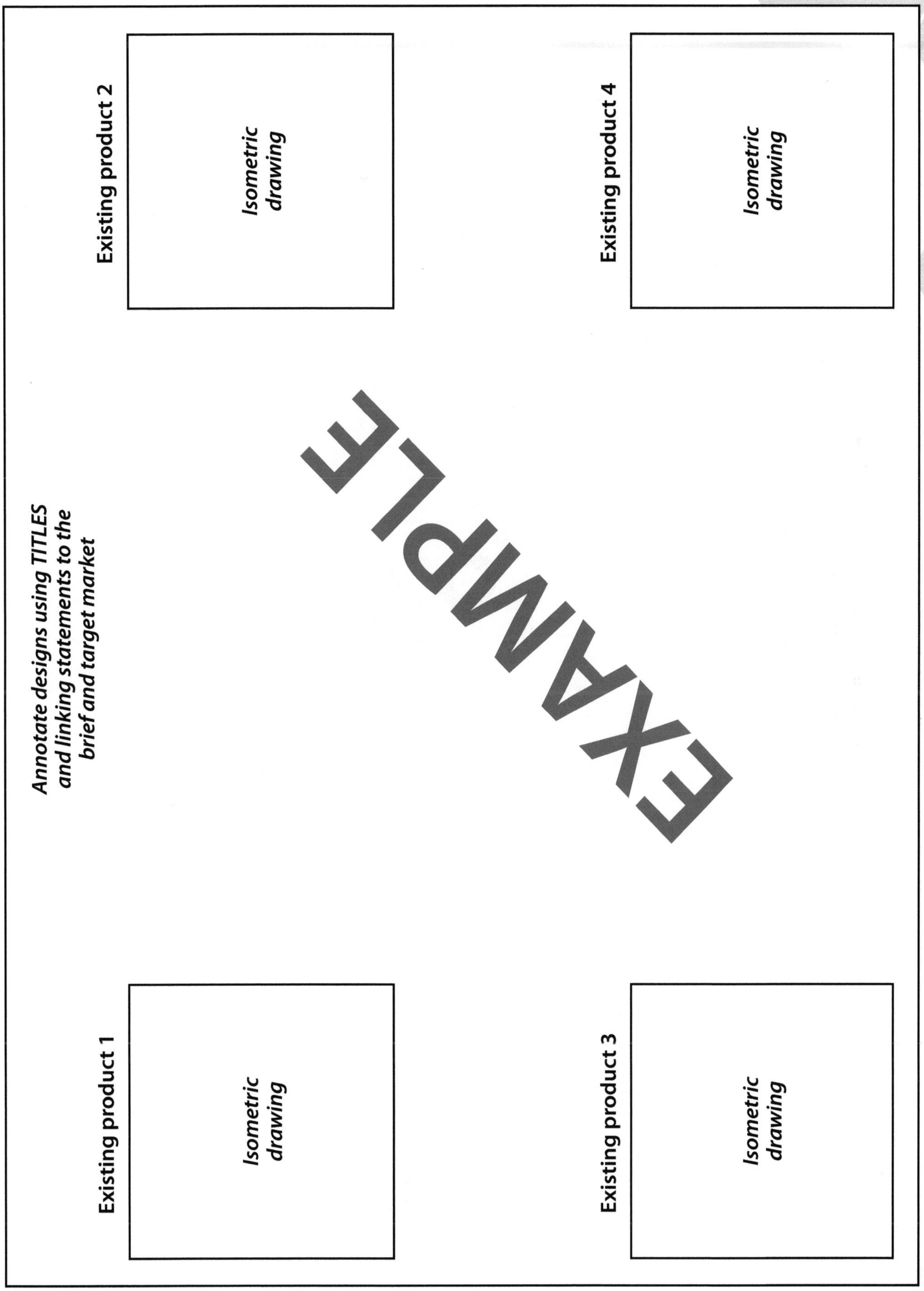

Annotate designs using TITLES and linking statements to the brief and target market

EXAMPLE

Existing product 1

Isometric drawing

Existing product 2

Isometric drawing

Existing product 3

Isometric drawing

Existing product 4

Isometric drawing

Reverse engineering

Use this page to reverse engineer a similar product to the one you will be designing. Split your analysis into EXTERNAL ANALYSIS and INTERNAL ANALYSIS. You can also use ACCESS FM titles.

EXAMPLE

External

Pictures of 'external' aspect of product

Internal

Pictures of 'internal' aspect of product

Design specification

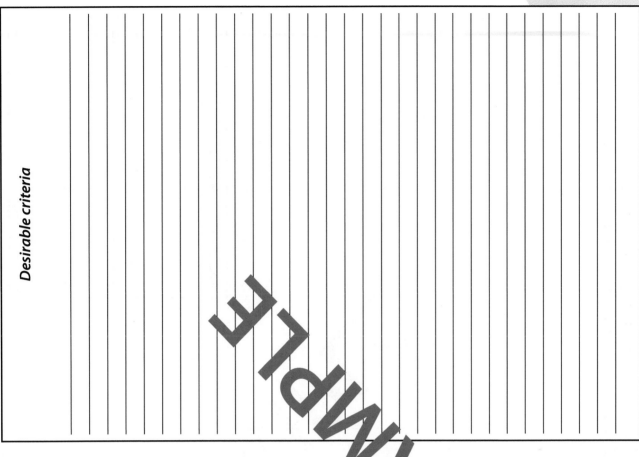

Desirable criteria

Essential criteria

EXAMPLE

Design solutions/initial ideas

Three or four isometric ideas created by you. Leave the CONSTRUCTION LINES on.
Annotate fully against the brief, specification and target market. Discuss materials and properties.

EXAMPLE

Final choice justification

Development of chosen idea

Three or four stages of development of your chosen design in isometric.

You must link AT LEAST two of your development stages to existing products (e.g. developed grip like a grip on a pair of pliers).

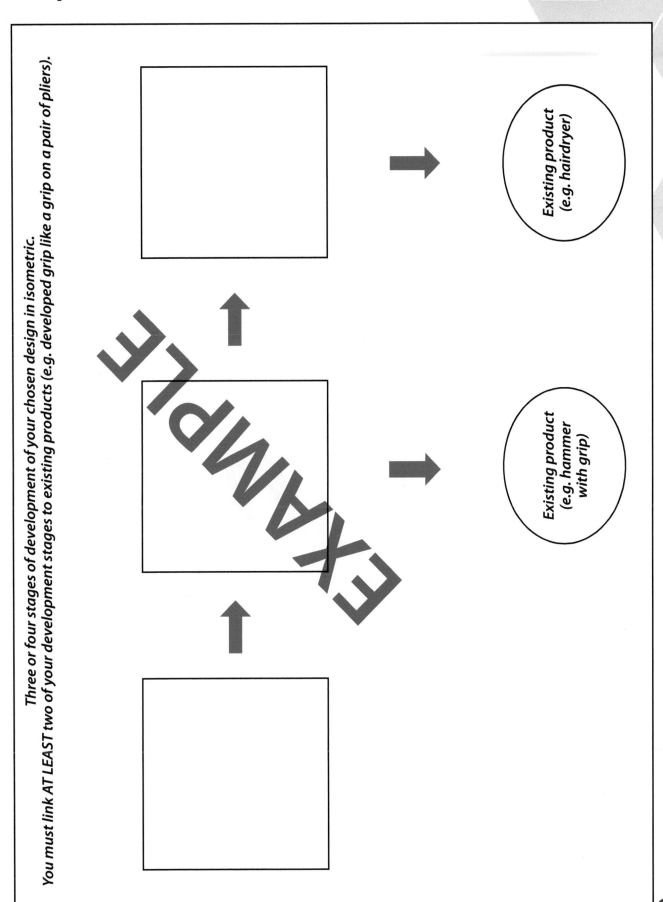

Existing product
(e.g. hairdryer)

Existing product
(e.g. hammer with grip)

EXAMPLE

Final solution

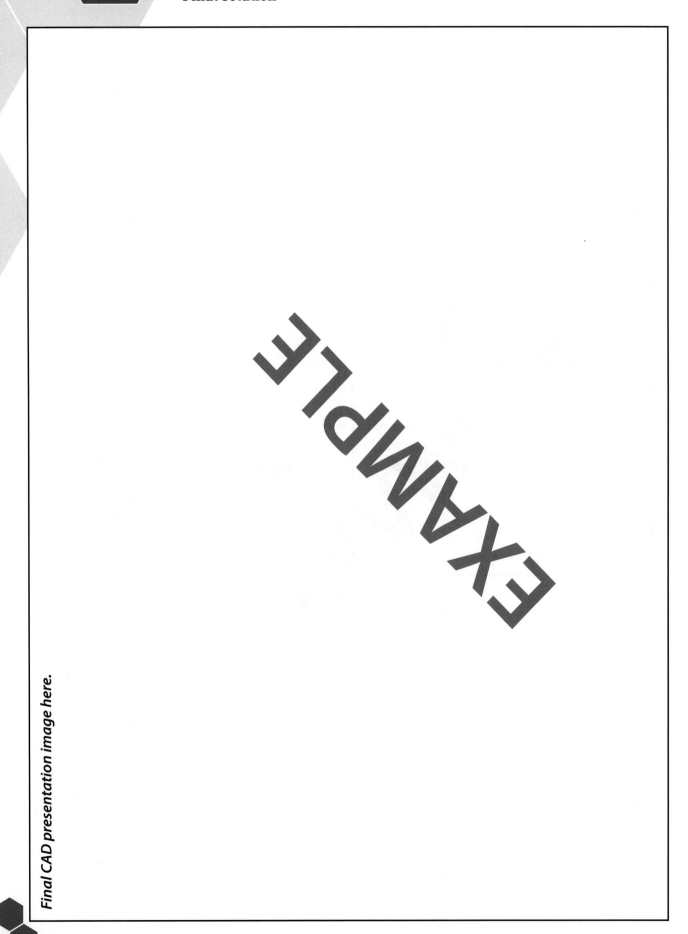

Final CAD presentation image here.

Engineering/working drawing

EXAMPLE

ALL MEASUREMENTS
IN mm

NAME:

DATE:

SCALE:

TITLE:

Mark schemes

Assessment criteria	Performance bands				Grade awarded
	Level 1 Pass	Level 2 Pass	Level 2 Merit	Level 2 Distinction	
AC1.1 Identify features that contribute to the primary function of engineered products	Identifies features that contribute to the function of engineered products although some features may not contribute to primary function.	Identifies accurately a limited range of features that contribute to the primary function of engineered products.	Identifies accurately a range of features that contribute to the primary function of engineered products.		
	Assessor comments				
AC1.2 Identify features of engineered products that meet requirements of a brief	Identifies features of engineered products although some features may not meet the requirements of a brief.	Identifies accurately a limited range of features that meet requirements of a brief.	Identifies accurately a range of features that meet requirements of a brief		
	Assessor comments				
AC1.3 Describe how engineered products function	Outlines how engineered products function with limited accuracy.	Describes how engineered products function.	Describes in some detail and some accuracy how a range of engineered products function.	Accurately describes in detail how a range of engineered products function.	
	Assessor comments				

Unit 2

Remember that the focus of the work could change but the task outcomes will be the same.

Novus Fabrication & Engineers Ltd – EXAMPLE BRIEF

Novus Fabrication & Engineers Ltd (NFE) manufactures engineering products and prototypes for product designer-based companies. A high street retailer has a design for a new desk lamp powered by an LED. The retailer has commissioned NFE to produce a prototype of the lamp before it starts full-scale production.

The retailer has provided NFE with all the relevant engineering drawings.

Task

1. Interpret the drawings and produce the prototype to scale.

2. Create a job/parts list that also records the tools, machines and health and safety equipment needed.

3. Create a 'sequence' or 'observation record' with photos of each process that can be followed by a third person, which also includes consideration of health and safety.

4. Evaluate the outcome of the lamp, including the working processes, tools used, accuracy, finish and health and safety.

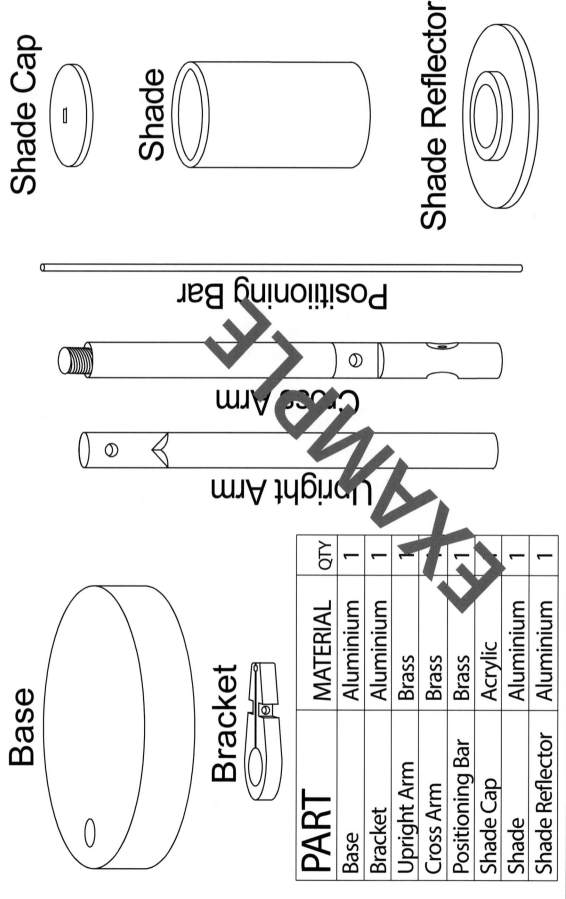

Shade Cap

Shade

Shade Reflector

Positioning Bar

Cross Arm

Upright Arm

Base

Bracket

PART	MATERIAL	QTY
Base	Aluminium	1
Bracket	Aluminium	1
Upright Arm	Brass	1
Cross Arm	Brass	1
Positioning Bar	Brass	1
Shade Cap	Acrylic	1
Shade	Aluminium	1
Shade Reflector	Aluminium	1

Title: Lamp Parts List

Materials: Aluminium, Brass, Acrylic

Client: Novus Fabrication & Engineers Ltd

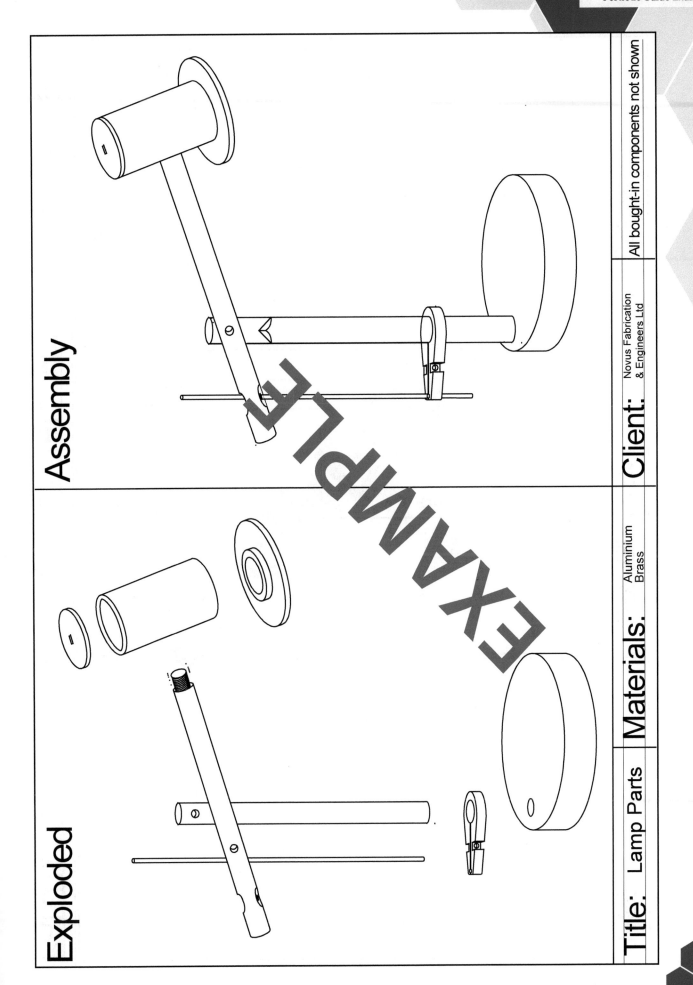

Assembly

Exploded

EXAMPLE

Client: Novus Fabrication & Engineers Ltd

All bought-in components not shown

Materials: Aluminium
Brass

Title: Lamp Parts

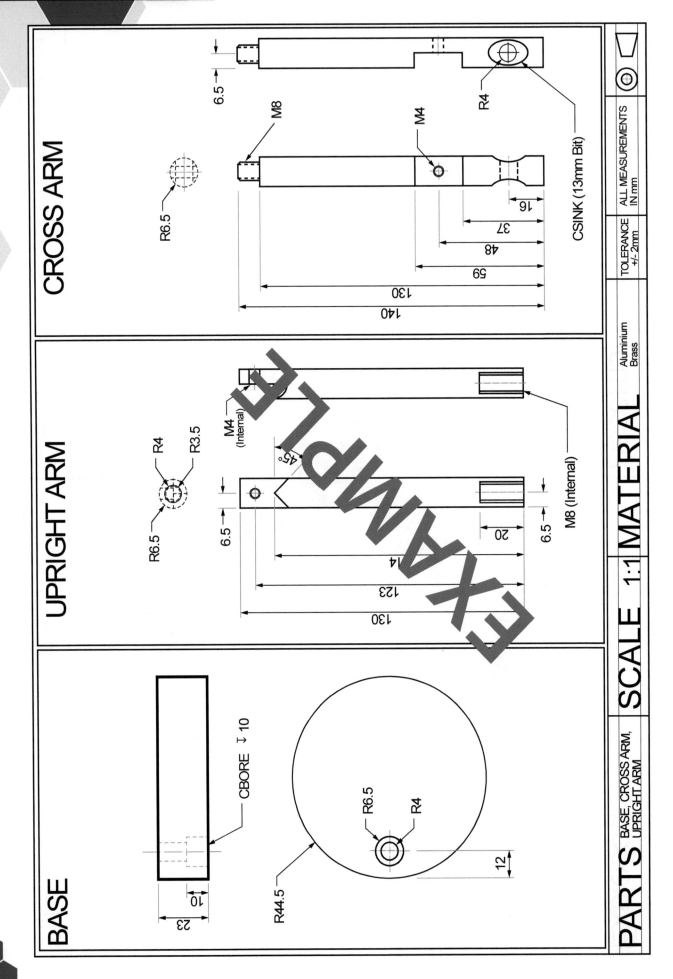

CROSS ARM

R6.5

M8

M4

R4

CSINK (13mm Bit)

16

37

48

59

130

140

UPRIGHT ARM

R4

R3.5

R6.5

M4 (Internal)

45°

M8 (Internal)

6.5

20

6.5

14

123

130

BASE

CBORE ⊥ 10

R6.5

R4

R44.5

12

10

23

EXAMPLE

PARTS — BASE, CROSS ARM, UPRIGHT ARM

SCALE 1:1

MATERIAL — Aluminium / Brass

TOLERANCE +/- 2mm

ALL MEASUREMENTS IN mm

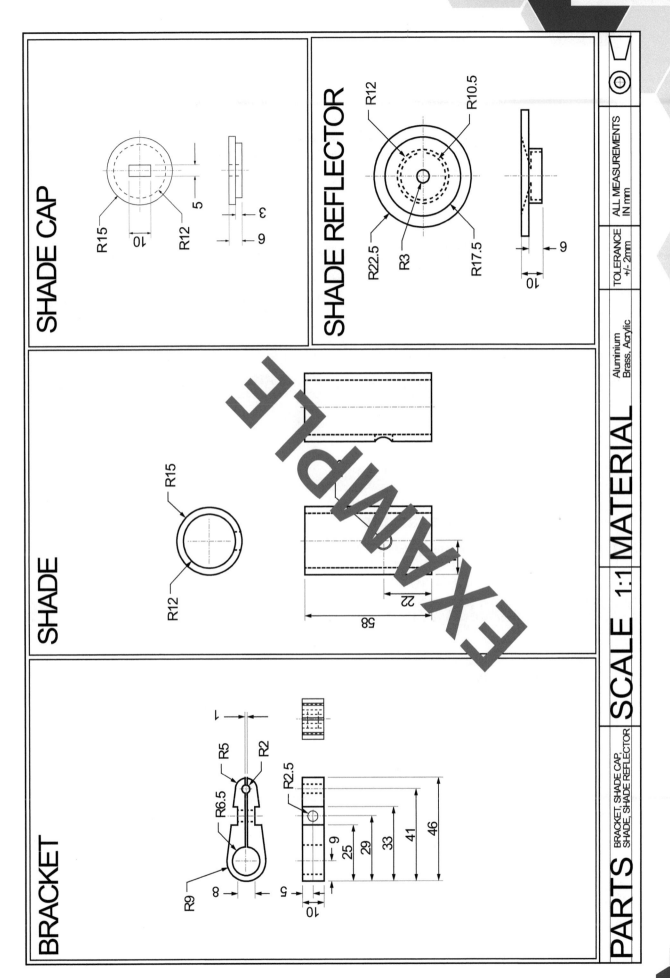

SHADE CAP

R15
R12
10
5
3
9

SHADE REFLECTOR

R12
R10.5
R22.5
R3
R17.5
9
10

SHADE

R15
R12
22
58

BRACKET

R5
R2
R6.5
R2.5
R9
9
25
29
33
41
46
8
5
10

EXAMPLE

PARTS — BRACKET, SHADE CAP, SHADE, SHADE REFLECTOR

SCALE 1:1

MATERIAL — Aluminium Brass, Acrylic

TOLERANCE +/- 2mm

ALL MEASUREMENTS IN mm

Job/part sheet pages example

You should be creating a series of tables explaining the **materials** and **equipment** needed to produce the prototype. The following table is an EXAMPLE ONLY of what you COULD produce. More than one table OR title would be needed if there is more than one part.

PART name/number			
Quantity	Material	Tools/machines needed	Health and safety equipment

PART name/number			
Quantity	Material	Tools/machines needed	Health and safety equipment

PART name/number			
Quantity	Material	Tools/machines needed	Health and safety equipment

GANTT example

You should create a GANTT chart that explains the **timings** of the making process. You must also list the TOOLS/EQUIPMENT that you are going to be using. Also, include QUALITY CHECKS (QCs) and HEALTH AND SAFETY (H&S) CHECKS as part of the chart.

Part: SHADE					
Student name:					
Process	Tools/ equipment	Quality check (*tick box*)	Health and safety check (*tick box*)	Time needed (*hours or 30-minute segments*)	

Sequencing/observation sheet pages example

You should be creating a series of tables explaining the **sequence of tasks** or **plan** needed to produce the prototype. The following table is an EXAMPLE ONLY of what you COULD produce. More than one table OR title would be needed if there is more than one part.

PART name/number							
Step/action number	Tool	Machine	Process	Health and safety	Time	Explanation	Photographic evidence

PART name/number							
Step/action number	Tool	Machine	Process	Health and safety	Time	Explanation	Photographic evidence

PART name/number							
Step/action number	Tool	Machine	Process	Health and safety	Time	Explanation	Photographic evidence

Evaluation sheet pages example

You should be creating an **evaluation** of the prototype production. The following titles may be used in any order you wish and even with your own translation of the wording.

This evaluation should have lots of reasoning. There should be words used such as:

- because
- as
- since.

You can evaluate each individual component or the project as a whole.

The evaluation should include:

Personal evaluation

- Use of tools and equipment
- Working to tolerance
- Health and safety
- Meeting deadlines.

Prototype outcome

- Accuracy
- Quality of finish
- Assembly.

Photographic evidence

- Component parts
- Assembled prototype (with different views).

179

Mark schemes

Assessment criteria	Performance bands				Grade awarded
	Level 1 Pass	Level 2 Pass	Level 2 Merit	Level 2 Distinction	
AC1.1 Interpret engineering drawings	Interprets limited information from engineering drawings with limited accuracy. Some information may not be appropriate.	Interprets information from engineering drawings with some accuracy. Some information may not be appropriate.	Accurately interprets most appropriate information from engineering drawings.	Accurately interprets a wide range of appropriate information from engineering drawings.	
	Assessor comments				
AC1.2 Interpret engineering information	Interprets engineering information with limited accuracy. Some information may not be appropriate.	Interprets appropriate engineering information with some accuracy.	Accurately interprets appropriate engineering information.		
	Assessor comments				
AC2.1 Identify resources required	A limited range of appropriate resources is identified. There are some significant inaccuracies and omissions.	A range of appropriate resources is accurately identified. There are some inaccuracies and minor omissions.	A range of appropriate resources is accurately identified.		
	Assessor comments				
AC2.2 Sequence required activities	A limited range of appropriate activities is identified. There is some attempt to sequence activities although not always taking account of external factors.	A range of appropriate activities is identified. There is some logical sequencing of activities, with some account of external factors.	A range of appropriate activities is identified. Most are logically sequenced, with clear account taken of some external factors.	Appropriate activities are identified and sequenced logically, taking clear account of a range of external factors.	
	Assessor comments				

Assessment criteria	Performance bands				Grade awarded
	Level 1 Pass	Level 2 Pass	Level 2 Merit	Level 2 Distinction	
AC3.1 Use tools in production of engineering products	A limited range of tools is used in engineering production. There is some evidence of safe working, although some intervention is required. The learner is able to access information or use tools with guidance. Use of tools may lead to a limited range of outcomes.	A range of tools is used in engineering production. There is evidence of independent safe working although some intervention may be required. The learner is able to use information or tools with limited guidance. Use of tools may lead to outcomes with some quality issues.	A range of tools is used effectively in engineering production. There is evidence of independent, safe working. Use of tools may lead to outcomes meeting most quality requirements.	A range of tools is used effectively in engineering production. There is evidence of independent, safe working. Use of tools will lead to outcomes meeting all quality requirements.	
	Assessor comments				
AC3.2 Use equipment in production of engineering products	A limited range of equipment is used in engineering production. There is some evidence of safe working although some intervention may be required. The learner is able to access information or use equipment with guidance. Use of equipment may lead to a limited range of outcomes.	A range of equipment is used in engineering production. There is evidence of independent safe working although some intervention may be required. The learner is able to use information or equipment with limited guidance. Use of equipment may lead to outcomes with some quality issues.	A range of equipment is used effectively in engineering production. There is evidence of independent, safe working. Use of equipment may lead to outcomes meeting most quality requirements.	A range of equipment is used effectively in engineering production. There is evidence of independent, safe working. Use of equipment will lead to outcomes meeting all quality requirements.	
	Assessor comments				

(continued overleaf)

Assessment criteria	Performance bands *continued*				Grade awarded
	Level 1 Pass	Level 2 Pass	Level 2 Merit	Level 2 Distinction	
AC4.1 Use engineering processes in production of engineered products	A limited range of processes is used in engineering production. There is some evidence of safe working, although some intervention may be required. The learner is able to access information or use processes with guidance. Use of processes may lead to a limited range of outcomes.	A range of processes is used in engineering production. There is evidence of independent, safe working, although some intervention may be required. The learner is able to use information or processes with limited guidance. Use of processes may lead to outcomes with some quality issues.	A range of processes is used effectively in engineering production. There is evidence of independent, safe working. Use of processes may lead to outcomes meeting most quality requirements.	A range of processes is used effectively in engineering production. There is evidence of independent, safe working. Use of processes will lead to outcomes meeting all quality requirements.	
	Assessor comments				
AC4.2 Evaluate quality of engineered products	Quality of engineered products is evaluated. Conclusions are mainly straightforward.	Quality of engineered products is evaluated using some appropriate techniques. Conclusions show some reasoning based on evidence.	Quality of engineered products is evaluated using mainly appropriate techniques. Conclusions show clear evidence based reasoning.		
	Assessor comments				

WJEC LEVEL 1/2 AWARDS IN ENGINEERING MARK RECORD SHEET

UNIT 2: PRODUCING ENGINEERED PRODUCTS

Learner name:

I confirm that the evidence submitted for assessment has been produced by me without any assistance beyond that allowed.

Signature: **Date:**

Assessor name:

The assignment brief used for summative assessment is attached, together with evidence of quality assurance.

I confirm that the evidence submitted by the learner has been produced under the controlled conditions set out in the qualification specification and model assignment.

The overall grade awarded for this unit is _____

Signature: **Date:**

Lead Assessor name:

I confirm that the evidence submitted by this learner for summative assessment has been quality assured and the grade awarded is confirmed as accurate.

Signature: **Date:**

Workshop Skills Projects

Bottle opener

The bottle-opener skills-based project on page 186 can include the processes:

- measuring

- scribing

- shaping – sawing, filing

- drilling

- finishing – painting, polishing, powder coating.

Door hook

The door hook skills-based project on page 187 can include the processes:

- measuring

- scribing

- shaping – sawing, filing

- drilling

- joining – pop-riveting, brazing, welding

- finishing – painting, polishing, powder coating.

Steel dice

The steel dice skills-based project on page 188 can include the processes:

- measuring

- scribing

- shaping – sawing, filing

- drilling

- facing off – centre lathe (four-jaw chuck to centre), vertical miller

- finishing – painting, polishing.

Fossil hammer

The fossil hammer skills-based project on page 189 can include the processes:

- measuring

- scribing

- shaping – sawing, filing

- drilling

- facing off – centre lathe (four-jaw chuck to centre),

- face milling - vertical miller

- creating internal and external threads (tap and die)

- sanding – linisher/belt sander

- finishing – painting, polishing.

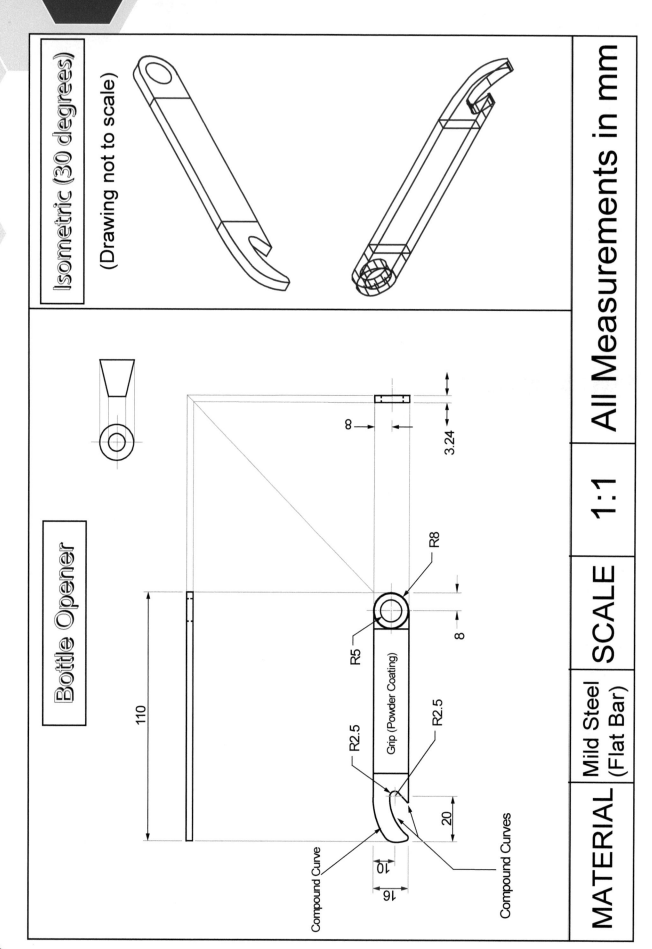

Isometric (30 degrees)

(Drawing not to scale)

Bottle Opener

110

R8

R5

Grip (Powder Coating)

R2.5

R2.5

20

10

16

8

3.24

8

Compound Curve

Compound Curves

| MATERIAL | Mild Steel (Flat Bar) | SCALE | 1:1 | All Measurements in mm |

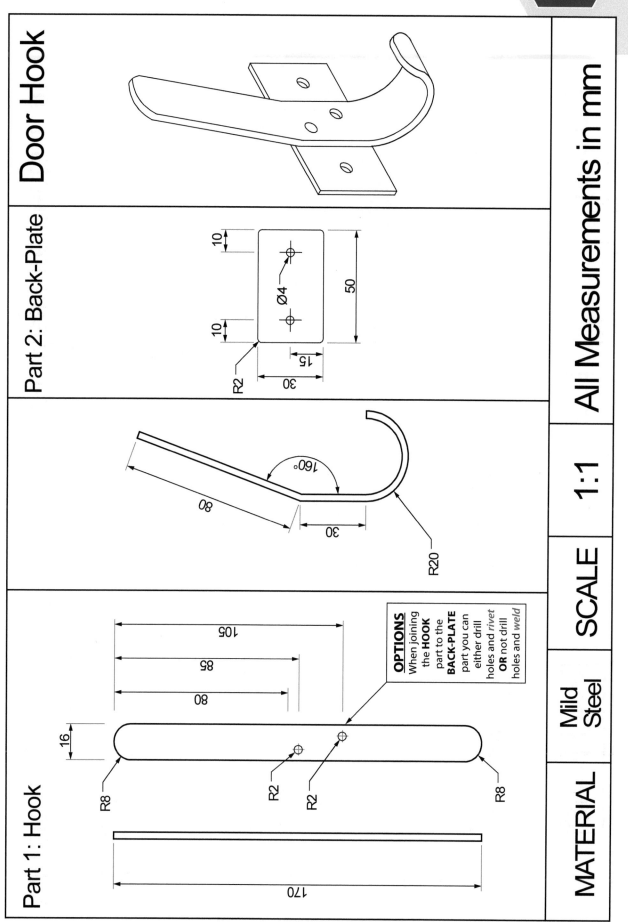

Door Hook

Part 1: Hook

Part 2: Back-Plate

OPTIONS
When joining the HOOK part to the BACK-PLATE part you can either drill holes and *rivet* OR not drill holes and *weld*

MATERIAL	SCALE		
Mild Steel	1:1	All Measurements in mm	

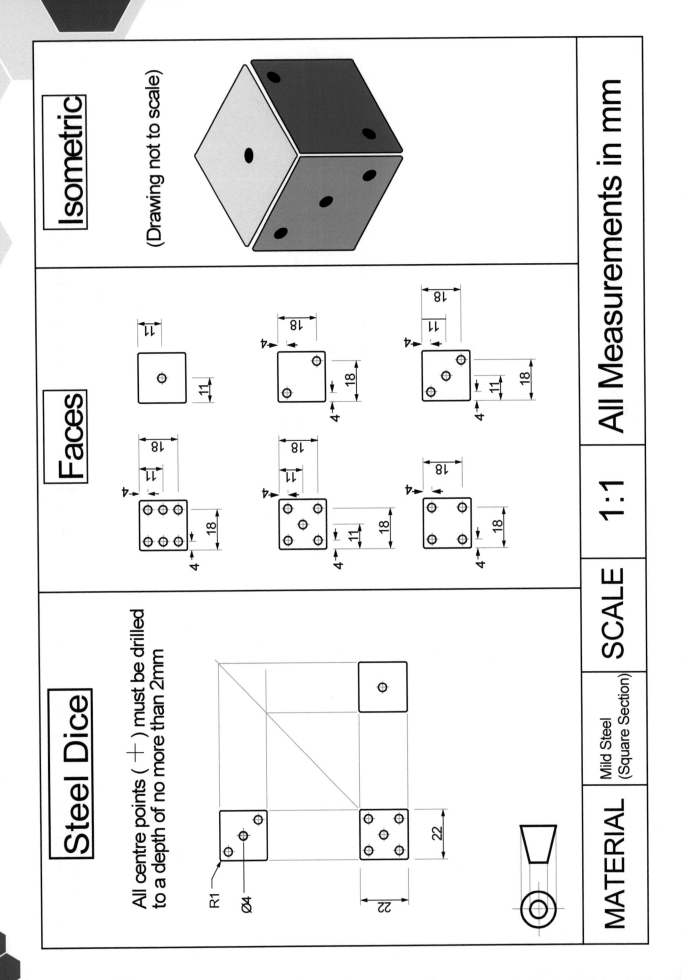

Isometric

(Drawing not to scale)

Faces

Steel Dice

All centre points (+) must be drilled to a depth of no more than 2mm

R1

Ø4

22

22

| MATERIAL | Mild Steel (Square Section) | SCALE | 1:1 | All Measurements in mm |

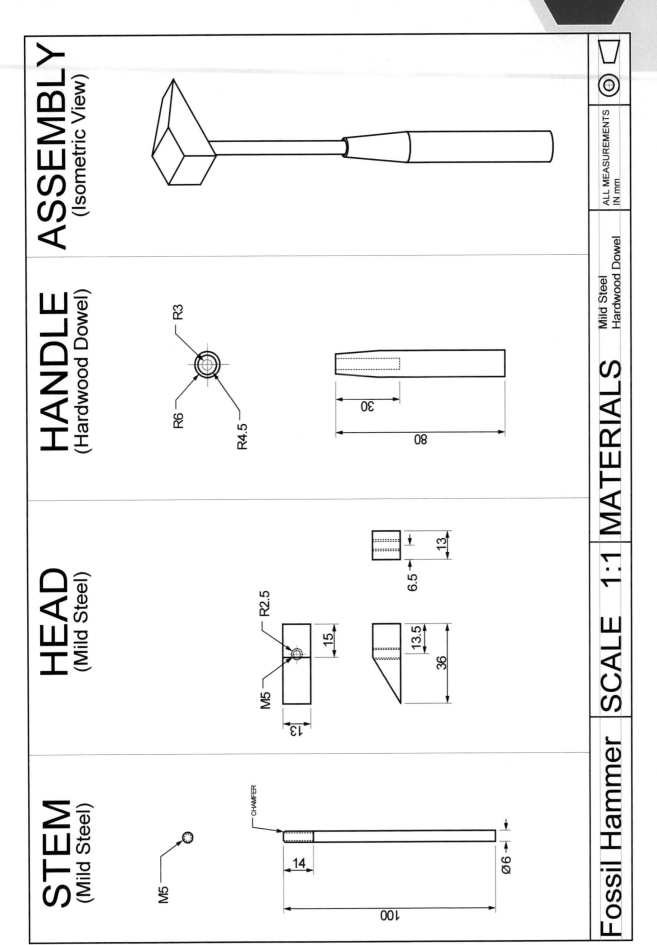

STEM
(Mild Steel)

M5

CHAMFER

14

Ø6

100

HEAD
(Mild Steel)

R2.5

M5

15

13

13.5

36

6.5

13

HANDLE
(Hardwood Dowel)

R3

R6

R4.5

30

80

ASSEMBLY
(Isometric View)

ALL MEASUREMENTS
IN mm

Fossil Hammer   SCALE   1:1   MATERIALS

Mild Steel
Hardwood Dowel

189

Glossary of Key Terms

Aesthetic: the way something looks to the eye.

Aqueduct: a bridge that carries water from one place to another.

Assembled: put together.

Axis: the direction of travel from a fixed point (X, Y and Z axes). The plural is axes.

Axonometric perspective: a pictorial representation of a 3D object that is not a true view of how you would view it.

Baseline: the horizontal line you use to 'level' your set square.

Billet: (or billet of metal) a piece of metal of a certain size that would be shaped by the forging process.

Bought-in component: a component purchased from another manufacturing plant.

Capillary action: when liquid (in the case of brazing molten metal) flows through very narrow surfaces such as two touching pieces of steel.

Catastrophically fail: to be tested until it is broken and does not work any more.

Chasing: the act of re-cutting the thread with a tap or die to repair any damage such as cross-threaded threads, as well as cleaning up the thread if it is old, worn or dirty.

CNC machines: machines that are computer numerically controlled.

Composite: something made-up from several parts or elements.

Compression: being squashed.

Condensed: reduced so anything not required is taken out.

Constraint: limitation.

Construction lines: faint, thin lines that are easy to rub out.

Conventions: technical terms.

Corrosion: oxidisation of a metallic surface or rust.

COSHH: **C**are of **S**ubstances **H**azardous to **H**ealth.

Crate: the 3D 'box' you start your isometric drawings with.

Criteria: specific headings or titles.

Cutaway drawings: designed to show the viewer important parts of the interior of an opaque object or product.

Electrode: an electrical conductor that is generally used to make contact with a non-metallic part of a circuit (e.g. the use of an electrode in welding, both mig and arc, where the electrode is the 'sacrificial' metal wire or rod).

Ellipse: a circle viewed in axonometric projection.

Expendable mould: a temporary mould that will be destroyed when the casting process is complete. It is not re-usable.

Exploded drawing: a drawing showing the parts of an object or product that have been separated to show how they interrelate or go together.

Extension lines: also known as lead lines.

Extrusions: profiles that have been extended or stretched.

Fabricate: to shape and join materials to create a product.

FEA: finite element analysis. A way to test your material (element) choice under forces on a computer model.

Feasible: manageable/possible.

Fillets: corner curves.

Footprint: the area of land each building takes up.

Force: the push or pull on an object causing it to change velocity (to accelerate).

FoS: factor of safety. To build in a safety margin when designing products.

Fossil fuels: non-renewable resources that can be burned to create energies (e.g. coal, gas and oil).

GRP: glass reinforced plastic; also called fibreglass. A mix of fibreglass and epoxy resin.

Hatching: a series of 45° parallel lines that are separated by an appropriate distance (e.g. 4mm) to show where a solid object has been cut.

HSS: high-speed steel. A high carbon steel that is very hard. Used for cutting tools such as drill bits and saw blades.

Innovation: from the Latin *innovare*, meaning 'to make new'. Take something that already exists and improve it.

Interrelate: work together.

Jacobs chuck: invented by Arthur Jacobs and patented in 1902; it is now a phrase used for the most commonly used chucks.

Jig: a work-holding device that can be used to create an operation repeatedly.

Key features: relevant pieces of information.

Kitemark: awarded by BSI when a product meets the standards.

LED: light-emitting diode. A diode that emits (gives-off) light.

Leyden jar: also known as a Leiden jar. A simple, glass jar with a foil-lined interior and exterior, used to store an electric charge – much like a battery.

Linisher: a machine that improves the flatness of a surface by sanding or polishing it. Also known as a linish grinder.

Malleable: pliable, easy to shape without breaking or cracking.

Mandrel: a cylindrical rod around which material is forged or shaped.

MDF: medium-density fibreboard. A man-made board used for many applications such as furniture and internal building projects.

Modelling foam: a foam material that is easy to shape with sandpaper. Also known as Styrofoam™.

Molten: a state of becoming liquid. Often refers to solid materials such as rock and metals that have been exposed to high temperatures and become liquid. Derives from the old English 'meltian', which means 'become liquid'.

Monomers: from Greek: *mono* = one; *mer* = part. Monomers are single molecules.

Obsolescence: a product is created to become obsolete – old-fashioned or out of date.

Optimal: the absolute best.

Orthographic projection: (in engineering) a means of representing different views of an object by projecting it onto a plane or surface.

Overall dimensions: the biggest dimensions.

Oxidisation: the process of oxidisation; where steel/iron surfaces react with the atmosphere and create ferric oxides (rust).

Pattern: a 3D copy of the item that is going to be cast. The sand is packed around the pattern and then the pattern is removed, creating a cavity that will be filled with molten metal.

PCB: printed circuit board. A circuit board that has been made using computer aided manufacture and is in fact 'printed'.

Perpendicular: something that is 90° to a given line (see figure on the left).

Perspective: your point-of-view when looking at an object.

Photons: light particles.

Plan view: a view from above; also known as a bird's-eye view.

Planes: the X, Y and Z axes (direction) you create in.

Platen: the part of a vacuum forming machine that acts as a shelf with holes and can be raised or lowered.

Polymerisation: the industrial process used to create plastics from naptha.

Polymers: a commonly known word for plastics. Polymers are chains of molecules.

Prism: a solid shape that has two ends of the same shape and size. The length can vary.

Product analysis: looking, feeling and maybe using the product to see how it works.

PVC: polyvinyl chloride.

Renewable energies: energies that can be renewed naturally such as wind, solar, geothermal and tidal.

Representations: views.

Research: the process of finding things out.

Resistor: an electrical component that can be used in a circuit to reduce/slow-down the current in it.

RPM: revolutions per minute; how fast the machine spins.

Runners: a channel that guides the molten metal to the part (or cavity of the part that will be filled with molten metal). A runner is considered waste material in relation to the cast product but can be re-used.

Scribe: mark-out.

Semiconductor:
- High conductivity = conductor (e.g. metals)
- Intermediate conductivity = semiconductor (e.g. silicon)
- Low conductivity = insulator (e.g. plastics).

Sequencing: putting tasks into the correct order.

Slag: waste material that is left when smelting or refining ore from metals.

SMA: shape memory alloy. A metallic alloy with a 'memory'.

Sprue: a hollow channel where molten metal is poured (or plastic is pushed). A sprue is considered waste material in relation to the cast product but can be re-used.

Stock-form: the size and shape the materials come in from the suppliers.

Swarf: a small piece of metal.

Sustainable: can the manufacture of the product be maintained (is it made from renewable resources)?

SWOT: stands for:
- Strengths: identifies good points about your project/idea.
- Weaknesses: identifies areas of your project/idea that could make it fail.
- Opportunities: identifies areas of your project/idea that you could exploit.
- Threats: identifies areas that could cause trouble for the project/idea.

Tailstock: the part of the centre lathe that sits towards the rear of the machine and holds various useful tools such as chucks, drill bits and centre guides, and can also support longer workpieces to stop them from wobbling when they are being rotated.

Tap wrench: a device for holding the tap. It also has 'arms' that have a textured surface for grip and can be rotated by hand.

Target market: the group of users you will be designing for.

Technical drawings: the common term used for 3rd angle orthographic projections.

Tension: being squeezed.

Thermochromic: a substance that changes colour according to temperature. *thermo* = temperature, *chromic* = colour.

Tolerance: an allowable amount of variation of a specified quantity, especially in the dimensions of a machine or part.

Top-down construction: when constructing a building with basement floors you can complete the structure of the higher floors before excavating and constructing the lower floors.

TPI: teeth per inch. How many 'teeth' a saw blade has per inch.

Trammel: (or the trammel method) uses a trammel of Archimedes (also known as an ellipsograph) to draw ellipses. This method can also be replicated by using a piece of paper and a major and minor axis of the ellipse to be drawn.

UPVC: unplasticised polyvinyl chloride. A hard form of PVC.

Urea formaldehyde polymer: a hard, slightly brittle plastic used for electrical casing/housing.

Vanishing points: lines that disappear into the distance.

Weighted lines: define the object you are drawing making it easier to see which lines to keep and which to erase.

Wet and dry paper: abrasive sheets of paper (such as sandpaper) for use mainly with metallic surfaces. It can be used either dry or wet. When wetted the moisture acts as a lubricant and removes the particles more quickly than when dry, creating a smoother surface, faster. Like abrasive tools and equipment (files, grinding wheels, sandpaper), wet and dry comes in different grades (grit sizes) for either a rougher or smoother finish. The lower the number/grit size of the wet and dry paper (e.g. 40 grit) the coarser the particles are on the paper and the rougher the finish, whereas the higher the grit size (e.g. 1,000 grit) the finer the particles are on the paper and the smoother the finish.

Index

Photo Acknowledgements

p. 1 pkproject / Shutterstock; p. 8 Gorodenkoff; p. 9 (top) Courtesy ISO; p. 9 (middle two) Courtesy BSI; p. 9 (bottom) Garsya; p. 11 (top left) tandemich; p. 11 (top middle) Kiryanov Aratem; p. 11 (top right) Age of Empires; p. 13 Steinar; p. 25 Bravo Ferreira da Luz; p. 34 Howard Davies, ISG; p. 36 (top) Stanislav Palamar; p. 36 (bottom) EloPaint; p. 40 (both top) Pixabay; p. 40 (bottom left) Image by rawpixel from Pixabay; p. 40 (bottom right) Ddabarti CGI; p. 41 (left) Pixabay; p. 41 (right) Photo Oz; p. 35 Artistdesign29; p. 37 fizkes' p. 38 cherezoff; p. 43 (top) Public domain; p. 43 (bottom left to right), Infinity Eternity, gresei, jemastock; p. 44 zhang Shen; p. 45 (top to bottom) robtek, jocic, Bildagentur Zoonar GmbH; Pavel_Klimenko; pinktree; p. 46 (top to bottom) Somchai Som, AlexLMX; PERLA BERANT WILDER; Teena137; Sideways Design; Juan Aunion; p. 46 (bottom left) chukov; p. 46 (bottom right) Jaromir Chalabala; p. 47 (top to bottom) tr3gin, Zanna Pesnina, Africa Studio, AnyVidStudio; p. 50 (top) Snoopy63; p. 50 (middle) Di Studio; p. 50 (bottom) Sue C; p. 51 (top) Paul Broadbent; p. 51 (bottom) Izf; p. 52 (right) Craig Russell; p. 52 (left) KreativKolors; p. 53 (top) Vaderluck at the English language Wikipedia; p. 53 (middle) TASER; p. 53 (middle right) faboi; p. 54 (top left) Mile Atanasov; p. 54 (top right) xiaoruil; p. 54 (middle) claudia veja images; p. 54 (bottom left) Epitavi; p. 54 (bottom right) Paul O'Dowd; p. 55 (top) Radu Crahmaliuc blog; p. 56 (top left) Aleksandr Medyna; p. 56 (top right) Incomible; p. 56 (middle) Kklikov; p. 56 (bottom) Avigator Fortuner; p. 57 (top) Shablon; p. 57 (middle left) Africa Studio; p. 57 (middle right) StockphotoVideo; p57 (bottom left) NinaMalyna; p. 57 (bottom right) sergey0506; p. 58 (top) Tu Olles; p. 58 (bottom) DmyTo; p. 60 (top) Es sarawuth; p. 60 (bottom) New Punisher; p. 61 Mooshny; p. 62 (top) PureSolution; p. 62 (middle) fad82; p. 62 (bottom) Igor Kardasov; p. 63 (top to bottom) fongbeerredhot, Mert Toker, MaxterDesign, Africa Studio, wk1003mike; p. 64 (top to bottom) Africa Studio, Victor Metalskly, fizkes, MIKHAIL GRACHIKOV, gualtiero boffi, 3D Vector, HM Design; p. 65 (top) 279photo Studio; p. 65 (bottom left to right) Beko, Dyson, Philips, Vytronix; p. 68 (top) Delices; p. 68 (bottom) SHEILA TERRY / SCIENCE PHOTO LIBRARY; p. 69 (top left) OlegSam; p. 69 (top middle) andregric; p. 69 (top right) Brovko Serhii; p. 69 (bottom) Arkadiy Chumakov; p. 70 goodluz; p. 72 nasirkhan; p. 73 Chaosamaran_Studio; p. 74 Sartori Studio; p. 76 (top) cobalt88, p. 75 (bottom) Andrii Yalanskyi; p. 76 (bottom) l I g h t p o e t; p. 78 magic pictures; p. 79 Mila Supinskaya Glashchenko; p. 80 create jobs 51; p. 83 Alexander Lysenko; p. 85 Mostovyi Sergii Igorevich; p. 93 mark_vyz; p. 94 A. and I. Kruk; p. 95 (top to bottom) Viktorija Reuta; petch one, Eleanor 3567, Safety Signs and Notices, Technicsorn Stocker, p. 96 (top to bottom) D Russell 78, Standard Studio, Janis Abolins, Alemon cz, Ave Na, Barry Barnes, Benjamin Marin Rubio, Technicsorn Stocker; Herry cai neng; p. 97 (warning signs) Jovanovic Dejan, p. 97 (prohibition signs) Krishnadas; p. 98 (all) Jovanovic Dejan; p. 99 Benjamin F Jones; p. 100 (top) AlexAMX; p. 100 (middle) Courtesy The Design and Technology Association; p. 100 (bottom) Courtesy CLEAPSS; p. 102 Andrey Eremin; p. 103 (top to bottom) Florian Schott / Creative commons, Facing off 6061 stock bar/YouTube, John F's Workshop, Florian Schott / Creative commons, Florian Schott / Creative commons, Pixel B, il2010; p. 104 (top) BUEngineer / Creative commons; p. 104 (middle left) Dmitry Kalinovsky; p. 104 (middle right) Armen Khachatryan; p104 (bottom) Vereshchagin Dmitry; p. 105 (top left) evkaz; p. 105 (top right) Eimantas Buzas; p. 105 (bottom) Rihardzz; p. 106 (top left) maksimee; p. 106 (top right) Steven Giles; p. 106 (middle left) By Glenn McKechnie – Own work, CC BY-SA 3.0, https://commons.wikimedia.org/w/index.php?curid=916098; p. 106 (middle right) By Glenn McKechnie – Own work, CC BY-SA 3.0, https://commons.wikimedia.org/w/index.php?curid=916098; p. 106 (bottom) exopixel; p. 107 (top left) Christopher Elwell; p. 107 (top middle) kariphoto; p. 107 (top right) Emrys 2 at en.Wikipedia / Creative commons; p. 107 (bottom left By Glenn McKechnie – Own work, CC BY-SA 3.0, https://commons.wikimedia.org/w/index.php?curid=916098; p. 107 (bottom right) Luigi Zanasi / Creative commons; p. 108 (top) By Glenn McKechnie – Own work, CC BY-SA 3.0, https://commons.wikimedia.org/w/index.php?curid=916098; p. 108 (middle) By Glenn